Winds of Change
Women in Northwest Commercial Fishing

Helen and Keane Gau's boat

WINDS
of
CHANGE

Women in Northwest Commercial Fishing

Charlene J. Allison,
Sue-Ellen Jacobs,
and Mary A. Porter

UNIVERSITY OF WASHINGTON PRESS

Seattle and London

Library of Congress Cataloging-in-Publication Data

Allison, Charlene J.
    Winds of change : women in Northwest commercial fishing / Charlene J. Allison, Sue–
Ellen Jacobs, and Mary A. Porter.
    p. cm.
    Bibliography: p.
    Includes index.
    ISBN 0-295-96840-0
    1. Women fishers—Northwest Coast of North America—Case studies. 2. Women fish
trade workers—Northwest Coast of North America—Case studies.   I. Jacobs, Sue–Ellen.
II. Porter, Mary A.   III. Title.
HD6073.F652N72   1989   89-16422
331.4'8392'09795—dc20   CIP

Title page illustration: Fishermen's Terminal, Seattle (*photo by Charlene Allison*)

Salmon-processing machine (*photo by Sonja O. Solland*)

*We dedicate this book to the many women
who shared their life stories with us.*

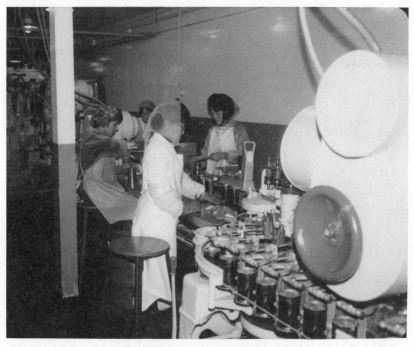

Fish-packing plant (*photo by Sonja O. Solland*)

Contents

Preface

In 1979 Michael Orbach, an anthropologist and administrator with Washington Sea Grant who had observed women fishing in various parts of the United States, suggested that Sea Grant would benefit from funding a research project detailing the roles of women in commercial fishing in the Puget Sound area. Sue-Ellen Jacobs developed the initial research design and wrote a proposal with the assistance of Sarah Jacobus, who left early in the project's implementation. Sue-Ellen Jacobs, then director of women studies at the University of Washington, and Marc Miller, assistant professor of marine studies at the University of Washington, were listed as coprincipal investigators for the study.

Six other people worked on the grant during the initial two years, interviewing women in commercial fishing, transcribing taped interviews, and maintaining the project's focus in biweekly meetings. These individuals included volunteers (Karen J. Blair, historian and assistant professor of women studies, and Judy Hodgson, historian and director of the Women's Information Center); paid interviewers (anthropologist Charlene J. Allison, and historian Leona J. Pollock); and administrative assistants (Jane Warner, and then Sara Stenson). Allison became the leading active researcher for the project and remained its principal organizer and scholar.

Over the course of four years of total work (including fifteen months of Sea Grant funding), numerous women and men in the fishing industry were interviewed. Between fall 1979 and spring 1981, eighteen oral histories of women working in fishing were collected by the project team. We have chosen ten of these life histories to represent a range of involvement by women in Pacific Northwest commercial fishing. These life histories were prepared from transcripts and taped interviews conducted by the following individuals: Charlene J. Allison (interviews with Christina Jefferson, Marti Castle); Leona J. Pollock (interviews with Marya Moses, Lois Engelson, Mars Jones, Helen Gau, Evie Hanson, Gladys Olson); Karen J. Blair (interview with Katherine "Tink" Mosness); and Marc Miller (interview with Linda Jones).

Although Sea Grant funding ran out before the collection of life histories was completed, Jacobs and Allison continued work with the research materials. In 1985 Joe G. Norman, Jr., associate dean for academic programs and research at the University of Washington Graduate School, approved two small grants for the project to hire Mary A. Porter, a graduate student in

anthropology. Porter then joined Allison and Jacobs in preparing and presenting the life histories. The authors are listed in order of the relative amount
of work each has put into the project over the years.

We wish to thank, in addition to those we interviewed, the following individuals: Stan Murphy, director of Sea Grant from 1969–80, for believing
in this project and providing the Sea Grant funding; Joe G. Norman, Jr., for
funding the final stages of research; Raleigh Watts and Loren Finley, who
know the sea, for assisting with technical information and the Glossary; Lloyd
Weller, at Everett Community College, for providing additional interview
materials on Marya Moses; the Washington State Department of Fisheries for
information on salmon-fishing regulations, the fish-receiving ticket system,
and the buy-back program; the Washington State Department of Fisheries
and Evergreen Legal Services for information on current legal developments
in Phase II of the Boldt Decision; the Alaska Commercial Fishing Entry
Commission for information on that state's limited-entry system; and the
National Oceanic Atmospheric Administration (NOAA), Seattle, for information on albacore regulations; Gretchen Swanzey for superb copyediting;
and Julidta Tarver and Naomi Pascal, our editors at the University of Washington Press; three anonymous reviewers for providing helpful comments on
the original manuscript; and Katherine (Tink) Mosness for writing the Postscript.

Introduction

This book provides a unique glimpse of a way of life characteristically associated with the Pacific Northwest. The stereotypic image of commercial fishing consists of male rugged individualists boldly confronting the sea. This stereotype obscures the range of actual behaviors and individuals who contribute to and make their living within the Pacific Northwest commercial fishing industry. The stereotypic image of individualists and the sea usually does not include women, who are often invisible in characterizations of these hardy souls. Nor does the stereotype include recognition of the effects on fishing activities of historical events and shifting political economies.

In contrast, our focus on women provides a view of the diversity of occupations and activities comprising a major industry of this geographic region and also a description of women's roles in that industry. Our task was to learn about the fishing industry while we learned about women's participation in its many facets. By bringing the diverse and seemingly disconnected lives of these women together and placing them in context of the fishing industry and its history we have come to understand the interconnectedness of their lives.

Before presenting the women's stories, our Introduction describes the goals of the project, our underlying assumptions, and our methods. Because we "as anthropologists believe that a knowledge of context is crucial for understanding" (Langness and Frank 1981:32), an overview of the industry plus a summary of significant events affecting Pacific Northwest commercial fishing over the past century have been included. In addition, most of the women did not present their stories in a strict chronological fashion and their stories cover periods of time which include significant historical events. We have imposed *time* on their stories both through the editing of the life histories (see Methodology, below) and the inclusion of a Time Line intercalating the year of each woman's entry into commercial fishing with 100 years of notable events affecting the industry. We have thereby attempted to place the women's lives and the events they discuss in the context of both the larger sweep of history and the passage of time. The reader is urged to review the Time Line before reading the life histories and to refer to it as needed to gain a sense of each woman's historical place. The reader is also referred to the Place-Name Locater (Appendix D) and to the Glossary.

In our Conclusion, we analyze the ten women's industry participation in terms of several situational and nonexclusive categories: fishermen's wives; women in small family businesses and independent women; fish processing; management; and political activism. Within this framework, we compare and contrast the women's work roles in terms of three important components of work activity—the nature of the tasks, and the social structural context and physical setting or locale in which the tasks are performed—in order to elucidate patterns in women's participation as demonstrated in the ten life histories.

Methodology

The goal of the Sea Grant project ("Sociocultural Roles of Women in Commercial Fishing in the Puget Sound Area") was to identify the range of roles women play within the fishing industry and to collect oral histories of women who have worked in the various roles. Because of the scarcity of social science information both on the fishing industry as a whole and on women's participation therein, we only suspected rather than knew that women's contribution to this industry is significant.

Our suspicions came from two major sources. Living in the Northwest, members of the project team had heard about women who worked in many components of the Pacific Northwest fishing industry. Furthermore, the growing body of literature on women's roles cross-culturally was demonstrating that women were more involved in all aspects of economic production than had previously been thought. We saw no reason why the same should not be true for commercial fishing. Among women mentioned in the local media were those involved in fishery politics, and we had all heard stories of women who fished in Alaskan waters or in the Strait of Juan de Fuca. Additionally, photographic displays showed a large number of women working in seafood canneries and processing plants. We wanted to know more about these and other women who might be found in the industry. We decided to use an approach directly derived from theoretical feminist anthropology of the 1970s. This approach assumed that women would be found in all sectors of the fishing industry, and that women would fish for all commercial species, using all appropriately associated types of gear.[1] If we found an area of the industry with no active women, we would ask why women were not now in that area and if they had ever been. Our bias was to expect women every-

---

1. We chose this positive approach rather than the negative approach used by Rosaldo (1974:17–42), Ortner (1974:66–87), and others. The proponents of the negative approach say that women everywhere are in one way or another subordinate to men and are devalued in the societies in which they live. They also seek to establish the relative powerlessness of women vis-à-vis men, and they argue that women do not participate in, or have access to, the domains in which men work.

where, and when we did not find them, to ask why. With this bias we un-
dertook a three-months' pilot study to establish contacts within the industry
and to determine the best way to document women's roles throughout the
industry.

One of our first activities was to interview attendees of Fish Expo,[2] an annual commercial fishing exhibition held alternately in Seattle and Boston
(see Appendix C for questionnaire). We asked respondents, among other
things, if they knew of any women involved in commercial fishing in any of
thirty-three different capacities. We found that 91 percent of our respondents knew of women who handle gear on commercial fishing boats, 55 percent knew of women who skipper commercial fishing boats, and 74 percent
knew of women who own commercial fishing boats.

Thus encouraged, project members contacted various commercial fishing
organizations such as the Puget Sound Gillnetters Association Auxiliary, Purse
Seine Women's Association, Pacific Coast Fishermen's Wives Coalition, and
many other organizations. Project members attended both informal and formal meetings of these organizations, gave presentations on the project, and
requested members' participation.

By the conclusion of the pilot study, we had learned that women are extensively involved in commercial fishing in a variety of capacities. We had
also learned that, although many fishermen and other industry workers reside
in the Puget Sound area,[3] they are very likely to work elsewhere. In short,
they are members of a broad occupational community rather than a geo-

2. Fish Expo is sponsored by a commercial-fishing journal, *National Fisherman*. The 1979 Fish
Expo was held in Seattle between October 24 and October 27, during the first month of our
project. *Alaska Fisherman* (November 1979) reported that 16,000 people attended the exhibition during the four days, including 2,200 from Alaska. We obtained ninety-nine completed
interviews using a short, open-ended, one-page survey. Interviewers, stationed at various locations in the exhibition area, soon learned to eliminate as possible respondents all who wore a
name-tag indicating they were not from the Pacific Coast of the United States or Canada. Other
possible respondents (such as gear manufacturers and suppliers) often proved to be unfamiliar
with the industry outside of their areas of specialization and so were unable to answer our questions. We completed interviews with sixty-three women and thirty-six men. Thirty of these
women (48%) fish, as do twenty-five of the men (69%). The highest percentage of respondents
were from the state of Washington (63%), and the second highest percentage were from Alaska
(16%). The remaining respondents were from other West Coast states and from British Columbia, Canada. Of our respondents 68 percent of the women and 75 percent of the men were age
thirty-nine or less.

3. The Puget Sound area was defined as the region which includes the following major bodies
of water (from south to north): Puget Sound, Strait of Juan de Fuca, San Juan archipelago, and
Georgia Strait (to the Canadian border). Although the name Puget Sound technically refers
only to the waters south of Port Townsend, Washington, the term is sometimes used to include
more northerly waters. For example, a 1979 Washington Department of Fisheries map showing
the Strait of Juan de Fuca, San Juan archipelago and Georgia Strait is labeled "Northern Puget
Sound Commercial Salmon Management and Catch Reporting Areas." A similar map showing
Puget Sound proper is called "Southern Puget Sound Commercial Salmon and Catch Reporting
Areas."

graphic community (the distinction between occupational and geographic communities is well illustrated in Pilcher 1972). Project members therefore decided to interview women resident in the Puget Sound area for at least part of the year, regardless of where their industry work was performed. Primary attention was centered on women fishermen,[4] processors, and women active in the politics of Pacific Northwest commercial fishing. Respondents were found through the Fish Expo interviews (the interviewer's last question had ascertained their willingness to be contacted again), fishing organizations, and personal networks. In addition, a five-minute report was broadcast on National Public Radio Station KUOW-FM in Seattle during the station's evening news program. Women who heard about the project, either through the broadcast or friends, contacted the project office about participating.

Had we had sufficient funds, we would have chosen to do a broad-based ethnographic study, using participant observation for at least one fishing season, but preferably for a year. In such a study, team members would have lived with and worked full time with members of the fishing community; made observational notes; recorded interviews among a relatively large number of people and in a variety of contexts; attended meetings where industry issues were debated and discussed; read the industry literature; and, in short, would have been totally immersed in the way of life known best to those who live it on a daily basis. Not having the necessary funds, we chose instead a route that would allow the voices of the women who have lived the fishing life to explore and explain the range of variations in work, politics, economics, and overall life-style of this critical aspect of the Northwest economy. In other words, we chose to collect tape-recorded life histories from a variety of individuals.

Oral histories alone could not give us the needed depth for fully understanding either the historical or sociocultural dimensions of women in commercial fishing. While oral histories are valuable sources of individuals' perceptions of their own lives and circumstances, they seldom provide sufficient information for generalizing to other individuals or situations. Oral life histories are often told in isolation from daily living; that is, people set aside time to tell their story to the researcher—the researcher does not have a chance to observe the life that is being described. We therefore provided contextual information for the life histories through attending industry meetings, conducting topical interviews with others in the community, and using library and other research sources such as photographs, periodicals, newspapers, books, research reports, and government documents.

In the months that followed our pilot study, project members talked with

---

4. At the time of the research, women who fished referred to themselves as "fishermen." Out of respect for their self-naming, we use this term throughout the book. Readers will note, however, that in the Postscript, Katherine Mosness has introduced a new term: "fisherwoman."

numerous women and men involved in the industry. We interviewed eighteen women in depth, using written consent forms (see Appendix A for consent form; see Preface for listing of who interviewed whom). For our taped life-history interviews, we chose topics designed to elicit information concerning each woman's experience in the industry (see Appendix B). With the exception of the interview with Linda Jones, all interviews were conducted in the women's homes or on their boats (sometimes these were the same). Each woman was interviewed several times, with the interviewer using each subsequent session to follow up on statements made during previous interviews. The interviews were tape-recorded, with informants' permission, and were later transcribed. Initial transcriptions were checked against the tapes and corrected.

We were interested in having the physical, spatial, and psychological environments conducive to reflexivity in presentation by the interviewees. We wanted the women to speak openly and candidly about growing up in their families and communities; the factors that had contributed to their decisions to become involved in fishing; their education; personal lives (including relationships with spouses, children, parents, and others); organizational involvement; views of the political context of fishing; personal values and expectations; impressions of the industry generally and about the future specifically; and special idiosyncratic or anecdotal aspects of their lives that they were willing to reveal. The variation in foci and emphases among the ten life histories is due to several factors: (1) the emphasis each interviewee chose in telling her own story; (2) the role or roles each individual played in the fishing industry; (3) the attentiveness of interviewers to subtle cues in the interview that led interviewers to pursue specific points (not all interviewees led interviewers in the same direction); (4) location and environment of the interview; and (5) circumstances, including age, under which the interviewees had entered the fishing industry.

Conducting oral history interviews and collecting contextual information allows researchers the opportunity to build stories of individuals' lives. Oral history researchers, in developing biographies, must be guided by an assumption that personal memories are often constructed by the interviewee to suit the perceived needs or desires of the interviewer (Ives 1980; Geiger 1986). It has also been found that individuals being interviewed for personal histories will reconstruct their lives in ways to emphasize either political or personal perspectives, or to provide a good image of the life they have lived (Langness and Frank 1981). These tendencies are sometimes uncovered when the researcher cannot fit what the interviewee has said to recorded historical fact or cannot corroborate interview materials about the same event or same period of time obtained from several individuals. When such disparities occur, the researcher can either talk with the interviewees again to ascertain that they have provided their "best memory" about the event or time period, or, failing the opportunity to re-interview informants, can present the recorded information *as* the best memory of the interviewees.

Each person in our study had read our statement of purpose (see Appendix A). Each was enthusiastic about documenting women's diverse experiences in the fishing industry. Each told her story (shaped by interviewer's questions) with attention to detail and an indication that she was providing her best memory of time and circumstances. With few exceptions, interviewees discussed dates and other matters without reference to documentation. Where these dates and events differed from those recorded in the media, legislative records, or other sources, we left the date as stated on the tape rather than alter the oral history materials.

In one instance, we combined taped statements collected by another researcher with the materials we had collected. Lloyd Weller had interviewed Marya Moses as part of a film project on Indian fishing. Incorporating portions of his tape added depth and richness to her life story.

Life-history research involves more than collecting oral histories and supporting documentation. To present individuals in publication, researchers must also engage in translation, interpretation, and analysis. In our case, *translation,* as that term is commonly used, was not required. Nonetheless, although all the women spoke to us in English, their use of fishing-industry jargon and concepts deriving from the fishing-industry culture meant that we had to learn the meanings they attached to specific expressions. For this reason, we have included the Glossary of specific fishing terminology and colloquial definitions for common terms. *Interpretation* occurred at three stages: in transcribing the taped oral interviews; in reorganizing these written statements to compose completed life histories; and in using the constructed life histories plus supplementary information as a basis for further analysis in our Conclusion. *Analysis* was conducted throughout every phase of our work.

Because of our need to convert raw tape-recorded interviews into sequentially structured life histories, we had to contend with the difficulties inherent in presenting the spoken word in written form. A conversational format had to be made readable: extraneous verbal hesitation forms and unrelated comments were eliminated and were replaced by ellipses (three periods); words were added in brackets (based on our understanding of meaning, or on statements made at other points in the transcripts or made off-tape and recorded in the interviewer's notes) to complete otherwise unfinished ideas and sentences.[5]

As our interviewees had not been required to tell their stories chronologically, sections of the interviews were rearranged. In structuring the presentation of the stories in what we perceived to be a logical and exemplifying

5. The original, complete transcripts are available to scholars, with controlled access, through the archives in Suzzallo Library at the University of Washington, Seattle.

manner, we interpreted experiences and statements in terms of social science concepts and the larger sociocultural domains of this region. In addition, historical events were sometimes used as pivots around which to organize a life history (e.g., Evie Hansen's "Before the Boldt Decision").

Once prepared, the life histories were sent to the women themselves for their editing (and as a check on the validity of our interpretations). Any section a woman wished to delete was dropped from her written life history, and pseudonyms were invented for those who requested this. The women's comments, changes, and other annotations (incorporated where desirable into the final versions) enabled us to be accountable to our informants and promoted maximum accuracy in our interpretations (cf. Watson and Watson-Franke 1985).

Our writing of the texts as they appear in the following chapters involved the following steps: (1) each of us chose several women's transcripts, constructing a time line and drafting a preliminary biography for each woman; (2) copies of the first drafts were distributed among us; (3) we met to discuss drafts, usually focusing on no more than one biography per meeting; (4) revisions of the first drafts were then distributed among us; (5) we met again and discussed the revisions. This process continued until we were satisfied that we had met our objectives. These objectives were to present a coherent story of each person's life, to highlight the interviewee's personal involvement in the industry, and to document a particular aspect of the industry.

The life-history texts are arranged in the book as part of the analytic process. The ten women have been chosen primarily because their lives illustrate a multiplicity of roles. The life story of a woman who fishes, is active in "fish politics," and is married to a fisherman, for example, provides information about three separate ways in which women may be involved in Pacific Northwest commercial fishing. We thank everyone who participated in the project. We hope that those women who were interviewed but whose life histories have not been included in this book will understand our point of view.

Pacific Northwest Commercial Fishing

Commercial fishing in the Pacific Northwest is a vast and complex industry. It is composed of at least five sectors: producers, processors, distributors, providers of support services, and fisheries managers and regulators. The producers are the fishermen themselves. Pacific Northwest fishermen pursue a variety of species using a number of different types of boats and gear. A single fisherman may pursue several species of fish, using different gear during the course of the year, and may fish over a very large area. Once the fish are caught, they must be cleaned to prevent rapid deterioration. Sometimes this is done on board by the fishermen; more often, the fish are delivered to a processing company where the fish are cleaned, then frozen or canned. Some companies buy the fish they process; other companies process fish belonging to yet other companies or individuals and store it for them. Once the fish have been processed, they must reach wholesalers and the retail market, a task performed by distributors. Distributors include everyone from independent fish brokers, who sell fish at retail or to restaurants and stores on behalf of particular fishermen or groups of fishermen, to large firms dealing in tons of fish. There are also firms that specialize solely in the storage and transshipment of fish from processor to wholesaler or retailer.

The producers, processors, and distributors all work toward a common goal: ensuring that the fish reach consumers. Providers of support services furnish the supplies and services necessary to keep those three sectors operating. Providers include boat builders, gear manufacturers, repair-service personnel, retailers selling clothing and other equipment specifically for fishermen, manufacturers of fish-processing equipment, and so forth.

The fifth sector of the industry—fisheries managers and regulators—governs the other four sectors. Their impact on producers is particularly significant. Managers and regulators determine which species of fish may be taken where, when, and by what type of gear. Fishing grounds for particular species are divided into regulatory areas. In June 1979, the Puget Sound region had thirty commercial salmon-management and catch-reporting areas. The areas were opened for fishing at different times and for varying periods of time during the overall season for a particular species. For example, area 6A (west of Whidbey Island) was most often open to nontreaty salmon fishermen (see *nontreaty, treaty* in Glossary; see also pages xxxiv–xxxvii below) of a particular gear type only one day per week between July 14, 1979, and September 18, 1979. During the week of August 5, reefnetters could fish for salmon in area 6A from 6:30 A.M. to 9:30 P.M. on Sunday; purse seiners from 5:00 A.M. to 9:30 P.M. on Monday; and gillnetters from 7:00 P.M. on Monday to 9:30 A.M. on Tuesday. A few weeks later, during the week of August 19, each gear type was allowed to fish for three or four days. Treaty fishermen faced similar regulations. Because the managers and regulators hold crucial power, the fishermen are often at odds with them.

Fisheries researchers might be thought of as a sixth sector of the industry, as their opinions and findings on the life cycles, habits, and numbers of the various species of fish heavily influence the activities of the managers and regulators. Many fishermen feel that the fisheries researchers are too influential.

Women in This Fishing Industry

Women fill active roles in each sector of the Pacific Northwest fishing industry, roles that often go unrecognized. Many women who are fishermen's wives are active in multiple sectors of the industry, and other women are involved in the industry in what might appear to be "noncustomary" occupations.

A common view of the fishing family assumes that men fish while women stay at home and take care of children and home work. Similar stereotypes are cited for workers in automobile factories, railways, and other occupations with a strong masculine image. The stereotype is based on several assumptions: that men can earn enough to support women and children, that they will provide that support, that married women occupy a leisure class when men are gainfully employed, and that all women will be married to gainfully employed men. But around the world, as Cook (1984) and others have found, unless they are members of economically secure families, most women will spend some portion of their lives working to support either themselves alone or themselves and others (e.g., their children, parents, and/or spouses), or because they *want* to work outside the home. To do this, they will enter any occupation, take any job available to them, or follow career paths for which they have been educated or trained.

Sometimes what is available is an extension of home work (e.g., domestic service, cooking) or an extension of family work (e.g., family farming). The fishing industry has many jobs that can be viewed as extensions of women's "customary home work": processing of food stuffs; financial management; supply purchasing; personnel management, which includes work assignment, overseeing, and selecting workers; planning and other administrative and practical work requiring judgement, decision making, delegation, organization, and more. The women whose lives we present here have engaged in these customary activities. At times, the available work has also included women's direct entry into the labor force in positions sometimes viewed as noncustomary. As discussed in our Conclusion, the distinction between customary or noncustomary work for women rests not only on the nature of the tasks performed but also on the social and physical contexts within which those tasks are embedded.

The term *fishermen* in the Pacific Northwest includes women who fish. The women identify with the occupation *fisherman*, yet they readily discuss gender issues that women encounter in the industry. Among women who fish alone or as captain of their own boats are Lois Engelson, Mars Jones, and Marya Moses. Women who work as crew members on boats operated by unrelated skippers, male relatives (husbands, fathers, brothers), or boyfriends include Helen Gau, Evie Hansen, and Christina Jefferson.

Women in processing most often work on the lines gutting and butchering fish, removing meat from shellfish, or packing and preparing fish for storage or distribution. Women have filled this role at least since the end of the nineteenth century (see Stevenson 1899: 516, plates 38 and 40), and their role as lineworkers is widely recognized. Women may also supervise a line operation and its workers or, on occasion, manage a processing plant. Christina Jefferson and Marti Castle exemplify these activities.

The women in distribution and marketing constitute a diverse group. Some are office managers or marketing representatives with large or small seafood companies. Others act independently. One woman we interviewed briefly, a retired fisherman (whose life history is not included here), worked for several years as an independent broker. She and another retired woman fisherman put together a buying operation in the San Juan Islands. Using a tugboat and two barges, they bought fish from fishermen, transported the fish, and sold them to restaurants and other consumers. Lois Engelson also became a fish buyer for a period of time after she retired from fishing.

Women are also providers of support services to fishermen and processors. They operate independent businesses providing gear manufacture and repair; for example, manufacturing black-cod and halibut gear and crab pots, and mending nets. They work in sales and other divisions of clothing and equipment stores. Women also monitor emergency radio channels and keep track of offshore weather conditions. In processing, women run businesses inspecting fish and certifying that they meet governmental export regulations.

In addition, fishermen's wives who do not regularly fish provide industry support services their husbands would otherwise have to perform themselves or hire someone to perform. Far from being solely "women who wait," these women often play active roles in their husbands' occupation. As noted by Cook (1984) and other scholars, unpaid women's work in industrial nations constitutes a hidden economic benefit that is often not included in the evaluation of gross national product and labor statistics. Wives may keep the financial records, help in regular boat and gear maintenance and repair, obtain emergency boat and gear parts for their husbands, and fill in as crew in emergencies. We include as critical actors in the industry those fishermen's wives who do not themselves go fishing, such as Gladys Olsen, who maintained her family home and took major responsibility for child rearing.

Within the last twenty years, many fishermen's wives have also acquired an additional role—political activist. Since their husbands are often away from long periods of time, the women keep track of political developments and lobby for or against new regulations. These activities can be seen in the life histories of Evie Hansen and Katherine (Tink) Mosness.

Finally, women may serve as managers and regulators with federal or state governments or with Native American groups, as can be seen in Linda Jones's account. There appear to be far fewer women in these positions than there are women who fish. Women also serve as observers for the National Maritime Fisheries Service on foreign fishing-processor ships.

There is considerable overlap in personnel between the five sectors. A fisherman may fish one season of the year and provide support services to other fishermen during the remainder of the year. A firm may process and ship fish, but not market it. Or, a company may own fishing boats, process its own fish, ship it, and also market it. There is also considerable interaction between personnel in the different sectors, both in the context of work and through personal networks. A woman who is married to a fisherman, but who does not herself fish, may be a marketing representative or may work in the office of a processing company. Fisheries managers and regulators constitute perhaps the most isolated sector of the industry. Even here, however, women (and men) who are politically active attempt to bridge the gap.

The life histories of the women in this volume illustrate this overlapping and interaction between the sectors. Although the emphasis is on producers and processors, activities typical of each sector are represented.

Arenas of Change

The Pacific Northwest commercial fishing industry has developed over the years through a complex interaction of events in many different arenas of change. The conditions created by these events have shaped the environment in which Pacific Northwest commercial fishermen live and work. The women in this volume discuss some of the major events; others are so familiar they go unmentioned. The following brief outline of circumstances and events shaping the industry provides a context for the life histories, for even passing comments take on extended meanings when their context is known.

The information presented here is necessarily abridged and is based on sources listed in the Bibliography. Any one of the topics discussed is suitable for lengthy study. Moreover, only developments significantly affecting those areas and species most important in the lives of the women presented here are included. Primary emphasis is on events affecting the Puget Sound area and the coastal waters of Washington State, and on salmon. Secondary emphasis is on events affecting Alaska generally, on albacore, and on king and tanner crab.

Species

All of the women in this volume have had some involvement in the harvesting or processing of salmon. Five species of salmon inhabit the northern Pacific Coast: chinook (or king), coho (or silver), sockeye (or red), pink (or humpback), and chum (or dog). Their life cycles range from two to six years, depending on the species. All are anadromous—they spawn in fresh-water streams, migrate to salt water where they mature, and return to their fresh-water streams of origin to spawn and then die. When they migrate to salt water, some salmon species stay close to the river mouth while others migrate thousands of miles. Some Columbia River chinook salmon migrate south to San Francisco Bay; others go north past Canada's Vancouver Island into the Gulf of Alaska. Some Puget Sound area chinook do likewise. Bristol Bay (Alaska) red salmon migrate yearly between Asia and Alaska, returning to Alaska to spawn at the end of their life cycle. Because of their migration patterns, salmon may be taken in rivers and bays as well as in various locations at sea. These patterns have extensive local and international implications, as will be shown below.

Many salmon fishermen also pursue albacore, a tuna found in the northern Pacific and usually caught 50 to 150 miles offshore. Albacore migrate between Japanese and North America waters each year. Early in the season, albacore returning to North American waters follow a southerly branch of one Pacific current and usually appear off Baja California in June. They then migrate northward. Albacore arriving later follow a second, more northerly branch and appear off the coasts of Oregon and Washington in early July. The albacore proceed northward and, as the coastal waters cool, go west to Hawaii and on to Japan. According to Browning (1974:11), "the time and place of appearance of albacore is determined solely by water temperature." In years of higher-than-normal water temperatures, albacore first appear north of Baja California and can range as far as the Gulf of Alaska.

King and tanner crab are found all along the Alaskan coast from Southeast Alaska to the northern Bering Sea. The major fisheries for these crab are centered at lower Cook Inlet and Kodiak Island in the Gulf of Alaska, in the Aleutian Islands, and in the Bering Sea. King crab live in deep water along the continental shelf until the breeding season, when they move into shallow water. The slow migration begins in late December or early January, with the crab reaching their breeding areas in March or April. In June, they begin to move once more into the deeper waters, where they remain for the rest of the year. There are several subspecies of king crab which, though they differ in size and color, have similar meat. The tanner crab, known in the marketplace as snow crab, is smaller than the king crab; it is found in Alaska in shallow waters.

Factors Affecting Availability of the Species

The abundance of each of the above species fluctuates with both natural and human-induced conditions. Some salmon have dominant cycle years when a great number of fish return to their home streams. In off years, much smaller numbers of fish may return. The cycle differs with each salmon species and place of origin. For example, Puget Sound area pink salmon return in sizeable numbers only in odd-numbered years; numerous Bristol Bay pinks return (following the important red salmon runs) only in even-numbered years. The abundance of albacore and of king and tanner crab is also subject to cyclical variations. According to Alverson and Pruter (1980:21), environmental factors linked with these changes include "availability of food at critical life history stages, changes in temperature patterns, predator-prey relationships, ocean currents, [and] upwelling processes."

Changes in water temperature are particularly important. Salmon and albacore alter their migration routes according to water temperature. Episodes of warming, due to a recurring climatic event in the equatorial Pacific Ocean known as El Niño, may occur as few as two or as many as ten years apart and may last eighteen to twenty-four months. In 1959, as a result of El Niño, albacore were caught in the Gulf of Alaska, far north of their usual range. During the 1982–83 season, the majority of Fraser River sockeye salmon abandoned their usual migration route through United States waters and were unavailable to American fishermen. This episode of El Niño had disastrous effects on salmon fishermen; in 1984 many of them applied for Small Business Administration disaster loans.

Human impact on fishing resources comes from a variety of sources in addition to the fishing activity itself. Salmon have certain environmental requirements that must be met if they are to reproduce. They need gravel beds in clean water within a certain temperature range in order to live and spawn. They must also be able to reach their spawning beds. The increasing human population in the Puget Sound area and elsewhere during the past hundred years has eliminated many salmon runs and has greatly reduced others through the alteration and destruction of salmon habitats. Industrial pollution, municipal waste, and adverse farming and grazing practices (which introduce pesticides and fertilizers) have lowered water quality. Sedimentation and silt caused by logging and mining practices, road building, and river channelization have smothered salmon gravel beds. Dams, log jams, and other obstacles have not only prevented salmon from reaching their spawning beds but have also altered the flow of water and the annual temperature cycles in rivers. Temperatures outside the range normal for salmon may initiate outbreaks of disease, and higher-than-normal temperatures may cause salmon to cease migrating or to die before spawning.

Development of the Fisheries

The most obvious impact on the abundance of fish is the level of human fishing activity. This activity has grown substantially over the years, aided by population growth and technological developments. This development of the fisheries, with particular attention to the effects of political events, is discussed below.

Salmon. When European and American explorers first met indigenous peoples of the Pacific Northwest, they encountered cultures centered around fishing as a subsistence economy. The earliest white settlers in this region pursued timber and trade industries, with farming, hunting, trading, and occasional fishing providing the main sources of their food. The commercial fishing industry as we know it today developed from early attempts to market salted fish. The Hudson's Bay Company packed salted salmon as early as 1823 on the Columbia River and later in several other locations in Washington and British Columbia, including their Nisqually post, established in 1833 on southern Puget Sound. American settlers also attempted to market salted fish, with limited success. Just one year after the founding of the city of Seattle in 1852, Doctor D. S. Maynard, a well-known figure in city history, packed almost one thousand barrels of salted salmon and fish oil and shipped them to San Francisco. Most of the salmon spoiled before reaching its destination. "Doc" Maynard and other settlers hired local Indians to supply them with the fish for their efforts. In fact, Indian fishing increased substantially in the 1840s and 1850s, to meet non-Indian demands for fish for local consumption and export and to provide Indians with the cash needed to obtain non-Indian goods. According to Bagley (1916:398), "most of the fishing was done by Indians, while the white men did the packing and selling of the product."

A major change in the fledgling industry came about with the introduction of salmon canning. The first West Coast cannery was established on California's Sacramento River in 1864. Canneries were soon found throughout the Pacific Northwest, and the first salmon cannery on Puget Sound was built in 1877.

The introduction of canning shifted the emphasis from salted fish and fish oil to a product form more conducive to large-scale distribution. Using canning, far more salmon could be preserved and could reach a larger number of consumers. The demand for fish therefore increased.

At about the same time, in the early 1880s, Puget Sound settlers began using fish traps as a primary method of catching fish. These traps were adapted from ones used by Northwest Coast Indians and consisted of man-made reefs constructed of pilings with nets attached to form a pathway or lead for the fish to follow into the "pot." Later on, floating traps, which were anchored in place, were devised as well. These traps became very profitable, and their owners often made fortunes. According to Bagley (1916:401), "the introduction of trap fishing stimulated the industry and resulted in a rapid increase in the number of canneries; also in the output of these plants."

The emphasis on traps, which were owned by a small number of individuals, usually packers, reduced Puget Sound Indians' participation in commercial fishing. Their participation was also reduced by a series of state enactments concerning licensing and permissible fishing times, gear, and locations. "By 1900, tribal fishermen lost their prime locations to huge trap nets strung between pilings and to state regulations which closed river fishing" (Droker 1979).

By the early 1900s, increasing numbers of immigrants arrived in the Puget Sound area. The area's early settlers had been from the Midwest or East Coast of the United States. The new arrivals were primarily ethnic Europeans who had fished in Europe. Some of these immigrants took up fishing for salmon with mobile gear of various types. The Slavs dominated purse seining, which became the economically dominant fishing technique in 1914 and 1915, and the Greeks dominated gillnetting. Immigrant Scandinavians concentrated on halibut rather than salmon. During World War I another gear type, trolling, appeared in great numbers.

Thus, by the 1920s several user groups were intensely interested in the salmon resource: Indians and original settlers, fishing primarily in shallow waters with fixed gear; and the European immigrants, fishing primarily with mobile marine gear. Another interest group was also gaining political strength—sportfishermen. Prior to 1925, steelhead, of prime interest to sportfishermen, had been included under state salmon regulations. In 1925 it was declared a game fish when taken in fresh water and could not be taken with nets. In 1933 sportfishermen successfully sponsored Initiative 62, establishing a separate state Department of Game with regulatory authority over steelhead.

Sportfishermen next combined forces with marine fishermen in actively supporting Initiative 77. Initiative 77 came about because, with the addition of mobile gear to the already efficient fish traps, the number of salmon began to decline. Marine and sportfishermen blamed the decline on the traps. Initiative 77 prohibited all non-Indian fishing with fixed gear such as traps, fish wheels, and set nets in Washington State. The initiative also divided northwest Washington waters into "inside" and "outside" areas, with different fisheries regulations applying on opposite sides of the dividing line. This distinction between "inside" and "outside" still exists and can be noted in the life histories in this volume. Washington voters passed Initiative 77 in 1934. According to Barsh, the banning of traps was essentially political:

> Trapmen were not only well-established in the fishery by 1900, they also represented the ethnic and social "brahmins" of Washington State society. Impoverished immigrants from seafaring nations found the salmon fishery already collected in the hands of an antagonistic ethnic establishment. . . . To the voters, the relevant question was who deserved to have fish [Barsh 1979:27–28].

In Alaska, too, the advent of canning allowed large-scale exploitation of the salmon resource. The Russians had exported small amounts of salted salmon from Alaska by 1830. The United States purchased Alaska in 1867, and by 1878 two American salmon canneries were in operation there. The number, location, and ownership of canneries in Alaska was largely dictated by Alaskan conditions at the time of initial American settlement. Because Alaska salmon runs occur over a large and isolated territory, and because highly efficient transportation and refrigeration techniques were lacking, processing operations were built close to the individual fishing grounds. Every season, the companies imported cannery workers and fishermen to Alaska, as the territory's resident population was sparse. Because of the expense involved in importing both fishermen and their gear and/or maintaining company-owned boats and gear, the canneries preferred fish traps as an efficient way to ensure a supply of salmon.

As in Washington, ownership of the traps was concentrated in the hands of a few. According to Cooley (1963:31), from 75 to 90 per cent of all traps in the Alaska fishery have been owned and operated by the canning companies, and 30 to 45 per cent of all traps were concentrated in the hands of five leading companies."

These canning companies were largely owned and operated by non-Alaskans. They also employed primarily non-Alaskans. Patterns that had developed during the very early years continued later to the detriment of Alaska's residents. As time went on, white and Native Alaskans were discriminated against; in some cases, they were not allowed to fish at all. Whereas other areas were given control of their fishery resources when they were organized into territories, in Alaska, through the influence of the large salmon packers, the federal government retained management of salmon from 1912 until one year after Alaska became a state in 1959. Alaskans felt they should be able to participate in their own salmon fishery and influence its management. Resource control and the elimination of fish traps, which were primarily in the hands of nonresident packers, were significant issues in the territory's decision to become a state. With statehood, control over the resource passed to the state. Despite this, antagonisms between resident and nonresident companies and fishermen still remain.

Albacore and Crab. The albacore fishery developed much later than the salmon fishery. Before the 1880s, albacore was considered a "trash" fish. It was first canned in 1906, but it was another twenty years before it came to be highly valued. Around 1900, the California albacore fleet consisted of bait boats,[6] with an influx of trollers between 1920 and 1925. Trolling has dominated the albacore fishery since that time, spreading to the Northwest in 1935. There has never been state or federal regulation of the albacore fishery.

The American king crab fishery developed only after World War II, and the tanner crab fishery even more recently. Prior to World War II, the Japanese had an extensive king crab fishery in Alaska and exported the product to the United States. This activity temporarily ended with the war. Americans had also caught and canned a very small amount of king crab during the

6. The operators of these boats release live bait-fish or "chum" among the albacore. The albacore go into a feeding frenzy and strike indiscriminately at the bait and at the fishermen's hooks.

1920s and 1930s, but the fishery did not receive much attention until the 1950s. A major change in the fishery came about with the introduction of crab pots as the principal harvest method, supplanting earlier use of tangle nets or trawl nets. In the early years of the fishery, boats constructed for a variety of other types of fishing were used in king crabbing, often with results disastrous for the crews of those boats. After 1965 new vessels designed specifically for Bering Sea king crabbing, fully equipped and costing over a million dollars each, were constructed. The first Alaska State closure of a king crab fishing season came in 1968.

Tanner crab began to receive attention in 1967. According to Browning (1974:22), "the difficulty of extracting the meat from the tanner crab had been one of the reasons for its neglect by Alaska crab fishermen." After the U.S. Bureau of Commercial Fisheries developed processing techniques and also obtained advice from the Japanese, the processing of tanner crab was made easier.

The Influence of Technology

The growth of the commercial fishing industry has depended in large part on the development of techniques and machines to process, preserve, and transport ever larger amounts of fish. The longer fish can be preserved, the greater the possibility of transporting it to distant markets and thereby reaching a larger number of consumers. The greater the number of consumers, the larger the amount of fish needed to meet the demand.

Technology in Processing and Transportation. A variety of machines and techniques to clean and butcher fish were developed to increase production. A machine referred to as the "Iron Chink" was introduced in a Bellingham (Washington) cannery in 1903 to clean salmon. It automatically cut off the head and fins, split the fish, gutted it, and cleaned it using jets of water. The machine gradually replaced large numbers of Chinese workers who had previously done all this work by hand. Modern salmon-cleaning machines have been developed from this early invention, but these can be used only for salmon. No similar machines are in general use for processing tuna. The mechanization of cleaning tuna has been hampered by the irregular shapes and sizes of the fish. Manual butchering of raw tuna is difficult due to the toughness of the fish; so the tuna are generally gutted, partially cooked, cooled, and then manually butchered. King and tanner crab are also largely processed by hand. The complete process is described in Christina Jefferson's life history.

Still other machines were developed to fill cans automatically. For salmon, machines fill cans of certain sizes while other size cans must be packed by hand. In tuna processing, a machine called the "pak-shaper" was introduced in 1948 to fill cans with a solid piece of fish in what is called "fancy pack."

Other machines fill cans with tuna in chunk or grated form. King and tanner crab are packed by hand.

Canning is one of the two most commonly used forms of preservation today. Other forms such as salting, drying, smoking, and pickling predate canning and freezing and are still used to a limited extent. Canning was first used in North America in 1841, although the real beginning of modern canning was in 1874 with the substitution of the pressure kettle, or retort, for boiling water. Canning was first applied to West Coast salmon in 1864.

Freezing is another very common method: in some cases, it supersedes canning. Not all fish, however, are suitable for freezing and frozen storage. Artificial freezing—natural freezing had already been used as a preservative technique—was introduced in Maine in 1861. By the beginning of this century, a number of different types of freezing equipment had been introduced. Today fish may be frozen by immersion in a medium such as brine, or, more usually, by direct contact with refrigerated coils (sharp freezing) or by air blast. One of the major hurdles to be overcome in the development of freezing as an effective technique was the prevention of moisture loss from frozen fish during storage and transport. This is now done through packaging, glazing, or a combination of the two. Christina Jefferson describes the process of glazing in her life history.

The developments of canning, refrigeration, and freezing have expanded the market for fishery products. Stevenson, writing at the end of the nineteenth century, says:

> The application of low temperature. . . . has enlarged and widened the general fishery trade so extensively that at present salmon fresh from the Columbia River, halibut from Alaskan waters, and oysters from Chesapeake Bay and Long Island Sound, are sold throughout the United States and in foreign countries, and numerous other fishery products are marketed thousands of miles from the source of supply, and for weeks after their capture, in condition not dissimilar to that when removed from the water [Stevenson 1899:358].

This expansion could not have taken place, however, without improvements in methods of transportation. The fish had to be moved from the fishing grounds to the processing plant and thence to the market, all in good condition.

In the early years of Pacific Northwest commercial fishing, salmon were caught and held alive in fish traps until needed, or fish were caught directly in the vicinity of the plant and delivered within a few hours. Once the technique of storing the catch on ice came into use in the early part of the century, boats could stay on the fishing grounds for longer periods of time. Crushed ice is still widely used, although some boats now have chilled brine or refrigerated sea-water systems. King crab boats must deliver their catch alive, and the crab are kept on board in holding tanks of circulating sea water. In recent years, air transportation has also been extensively used, especially in Alaska.

The first improvements in the transportation of fish to distant markets came with the development of refrigerated railroad cars cooled by crushed ice and rock salt in the 1860s and 1870s. Stevenson reports (1899:368) that "the shipment of fresh salmon in carload lots began in 1884, during which year eight carloads of fresh salmon were sent east [from The Dalles, Oregon], all arriving in good condition." Frozen salmon were delivered to the East Coast by the same method. The marketing of frozen fishery products, however, did not begin until 1926, when dry ice began to be used to cool the rail cars. The railroads soon began to experience competition from the trucking industry. The first refrigerated trucks were on the road in 1905, and trucks with mechanical refrigeration systems came into use in 1925. Modern-day equipment developed from these early beginnings. Today, trucks transport the largest share of seafood products in the continental United States.

More recently air transportation is beginning to challenge other forms of conveyance. It has been a primary method in Alaska for some time, as that state has few roads. One of the problems encountered in transporting seafood by air has been fluid leakage from seafood products. This fluid can cause severe corrosion and can damage the aircraft's operation system. To deal with this, leakproof containers have been developed, and only dry ice or "gel packs" (reusable plastic bags filled with a mixture of water and chemical compound) are acceptable as refrigerants. The role of air and other modern forms of transportation in the functioning of a processing plant is illustrated in Marti Castle's life history.

Technology in Harvesting. The capacity of the fleet to catch greater amounts of fish, and to venture farther offshore in search of fish, grew with the application of technological developments to commercial fishing. In the early days of the industry, different types of engines replaced sail and human power over a relatively short period of time. Steam-driven vessels were used in some fisheries in the 1890s, and gasoline engines suitable for smaller vessels were available by 1910, although earlier models had first appeared in 1903. In 1909 approximately 17 percent of Washington State's salmon vessels were under power, while the remainder were powered by sail, or by rowing (see Cobb 1911:69). In fact, Washington led other West Coast states in this conversion to power. In 1909, 13 percent of Oregon's salmon fleet and 3 percent each of Alaska's and California's were power vessels. Ten years later, almost all fisheries had converted to power. Diesel engines were the next improvement; by the 1930s they were replacing gasoline engines.

Over the years, various laborsaving devices have been incorporated into commercial fishing equipment. During the first half of this century, these devices were operated manually, via take-offs from the main engine or by electricity. By the end of the 1940s, for example, most trollers had gurdies, or large reels, to haul in their lines. These gurdies were and are usually run by some sort of power, although hand-operated gurdies still exist. Similarly, when seine drums (large rotating drums on a vessel's afterdeck used to haul in a seine net) were first introduced in 1950, they ran on mechanical power. These drums, which were first used on Puget Sound seiners, are so efficient that their use has been banned in Alaska.

The introduction of hydraulic power in the 1950s brought about a number of significant changes. Hydraulic power was applied to previously existing equipment, such as gurdies on trollers and the seine drum, as well as to entirely new devices. Hydraulic line haulers were installed on albacore trollers and were adapted for use in king crab fishing. Hydraulic blocks to pull crab pots replaced seine winches and various blocks and tackle. A new hydraulic device for seiners, the Puretic block, was introduced in 1955. The Puretic block was invented by Mario Puretic, a Californian, and was first manufactured by a Puget Sound firm. Since then, its use has spread around the world. The block resembles a very large pulley, and it hangs overhead suspended from the vessel's boom. The net runs through the block and, as it does, is lifted out of the water and onto the vessel's deck. Both the Puretic block and the hydraulic drum are used on contemporary seiners; some seiners even have both.

The application of hydraulic power has effected a reduction in crew size on many vessels. Hydraulic blocks on smaller king crab vessels have allowed skippers to reduce their crews from five or six to three or four; in seining, skippers who installed the hydraulic drum could reduce their crews from eight or ten to three or four, or to four or six with the Puretic block.

The 1950s also brought about a revolution in netting materials. Nylon netting was introduced in about 1954. Prior to that, nets were made of cotton. Cotton nets were heavy and had to be replaced every year or two at great expense. Nylon netting is lighter in weight and deteriorates less rapidly than cotton.

New electronic navigational aids appeared during the years following World War II. Prior to the war, most vessels had radio direction finders (RDF) and depth sounders as aids to navigation. An RDF can be used as a homing device or to obtain one's approximate position at sea. A depth sounder, a type of echo sounder, is used primarily to determine water depth and to avoid underwater obstructions. Although improved models of these devices are still in use and can be found on most vessels, the navigational aids most important to present-day commercial fishermen are radar and Loran. Radar (the term is derived from the words "radio detecting and ranging") was first developed in the late 1930s but was not available commercially until 1945. It gives the distance and relative bearing of objects, especially in close proximity to the boat, and is useful for navigating in coastal areas, narrow channels, and situations of low visibility. It is especially useful in the Pacific Northwest because of frequent rain and fog.

Loran, a long-range navigational system using pulsed signals transmitted by two pairs of shore stations, enables a skipper to obtain accurate bearings. The Loran signals cover an area of water fifty nautical miles out from shore, or to the hundred-fathom curve, whichever is farther. The original Loran system, Loran A, is being replaced by Loran C, an even more accurate version of the system. Wasserman quotes a fisherman on the use of Loran:

> Pre-Loran fishermen needed a trained eye. "You fished an edge with visual ranges," Jones recalls. "Yonder rock lined up with yonder mountain peak." When a certain rock lined up with a certain tree, it was time to change course because the edge was moving in another direction. "Electronics allows you to fish further offshore," Jones explains. "It always was a constant worry. Before you'd drop your gear overboard and never find it again. Now, there's no worry what-so-ever" [*The Fishermen's News* 1980, 36(20):32].

Besides serving as aids to navigation, echo sounders of various types help to locate fish. These devices can locate schools of fish at a distance, can distinguish between a single large fish, schools of fish, and the ocean bottom, and can determine the nature of the bottom. Some sounders send their sonic pulses straight down, while others are able to scan the water in all directions.

These and other new electronic equipment allow fishermen to fish farther offshore, monitor the ocean bottom to avoid losing gear or tearing nets, and find gear that has been previously set. In addition, improved communication equipment is also available. Fishermen may use VHF radio to communicate with shore, the U.S. Coast Guard, or one another, or to make a call on the VHF emergency channel. They are more likely to communicate with one another, however, through their citizen's band (CB) radios, which are often left on continuously. Lois Engelson's life history illustrates fishermen's use of CB radios.

The Politics of Scarcity

As the number of salmon decreases due to any or all of the reasons previously mentioned—natural conditions, pollution, adverse land use and development, increased fishing effort and efficiency—political maneuvering to assure access to the resource increases in intensity. On the international level, many countries such as the United States, Canada, Japan, the Soviet Union, and Korea have an interest in North Pacific fishery resources. On the local level, interested groups include American Indian treaty fishermen, nontreaty commercial fishermen, and sportfishermen. The various states and regulatory agencies also have the responsibility to ensure continuation of fishery resources. A share of the resources must therefore be allocated for conservation. We have already seen how conflicts between user groups arose early in the development of the fishing industry. Settlers edged out Indians, marine fishermen edged out those with fixed gear, and sportfishermen carved out a definite share for themselves. This maneuvering to retain or increase shares in the resource has continued and intensified both locally and internationally over the years.

This is particularly the case with the salmon resource. Because of salmon migration patterns, various countries must sometimes share the same salmon stocks. Events and fishing activities in one country often affect fishing in other countries. This became readily apparent in 1913, when a rock slide in the Hell's Gate Canyon in British Columbia, Canada, nearly wiped out all the Fraser River sockeye runs. Fraser River salmon constitute an important fishery for both U.S. and Canadian fishermen. After a long series of talks, the United States and Canada signed a treaty in 1937 to set up the International Pacific Salmon Fisheries Commission (IPSFC), composed of representatives of both nations, to manage and distribute Fraser River–spawned sockeye salmon. The commission restored the Fraser River sockeye runs and promulgated its first regulations in 1946. The commission's jurisdiction was later extended to include pink salmon from the Fraser River and various Puget Sound streams. Crutchfield and Pontecorvo note (1969:131) that when this happened in 1957, "a substantial proportion of the total catch of the Puget Sound area, by weight and value, was brought under international regulation."

Prior to the establishment of the twelve-mile limit in 1966 (see below, page xxxii), U.S. and Canadian fishermen fished up to three miles off each other's coasts. In 1970 the United States and Canada began signing reciprocal fishing agreements as to areas open to each other's fishermen and shares of the catch. As more and more was learned about salmon migration patterns, it became apparent that other runs besides those of the Fraser River and Puget Sound crossed international boundaries and passed through both United States and Canadian jurisdiction. These agreements were renewed or renegotiated periodically until 1978. For a number of years after that there were no reciprocal fisheries agreements between the two countries regarding salmon, other than the regulations of the International Pacific Salmon Fisheries Commission. A new salmon treaty was signed by the United States and Canada in March 1985, and the IPSFC was absorbed by a new and broader U.S.–Canada Salmon Commission on December 31 of that year.

In the 1930s antagonisms developed between the United States and other countries over the latter's exploitation of Alaskan fisheries. In particular, the Japanese had begun king crabbing and also trawling for groundfish in the eastern

Bering Sea. Foreign fishing ceased during World War II, but after the war there was an explosion of activity. In 1952 the Japanese began their first high-seas salmon fishery near the western Aleutian Islands. They also resumed king crabbing and other activities.

In the late 1950s the Soviet Union joined Japan in fishing off the U.S. Pacific Coast. They began fishing for king crab and flat fish in the eastern Bering Sea and later expanded to fishing for other species and in other areas. Whereas the Japanese concentrated their efforts in Alaska, the Soviets expanded to the coasts of British Columbia, Washington, Oregon, and northern California in the mid-1960s. In the 1960s the South Koreans also joined the Japanese and Soviets in fishing off Alaska. In particular, in 1969, the South Koreans began gillnetting for salmon of U.S. origin.

Prior to the Magnuson Fishery Conservation and Management Act of 1976, the United States attempted to control foreign fishing off its coast through a series of treaties and agreements with the various countries. For example, a treaty signed between the United States, Canada, and Japan went into force in 1953. Through the provisions of this treaty (International North Pacific Fisheries Convention), the Japanese agreed to abstain from fishing for halibut of North American origin, and for herring and salmon east of a boundary line designated at 175 degrees west. The Japanese reserved the right to fish for herring and salmon west of this line and in the Bering Sea.

The need to protect the fishery resource through more than fishing treaties and agreements became apparent to the United States government with the explosion of foreign fishing activity in the 1960s. According to Jacobson (1981:13), "international law recognizes that custom has established at least a 12-mile territorial sea." That is, each country claims absolute sovereignty over the area included within twelve miles of its coasts. The only exception to this sovereignty is that "other nations have the right of innocent passage" (Jacobson 1981:13). For military reasons, however, the United States has claimed only three miles as "territorial sea" and has insisted other nations do likewise. In 1966 the United States added a nine-mile zone, referred to as "the contiguous zone," to its three-mile territorial sea. The United States claimed the additional nine miles, not as part of its territorial sea, but as a zone in which to exercise exclusive fishing rights and limited enforcement measures, such as making smuggler arrests. This is the "twelve-mile limit" referred to in some of the life histories. Canada had extended its jurisdiction to include the additional nine miles in 1964.

By the late 1970s the United States had become very concerned about the effects of foreign fishing in general on United States fishing resources. In 1976 the 12-mile zone was extended to 200 nautical miles by the Magnuson Fishery Conservation and Management Act. The purpose of the act was "to conserve and manage all fishery resources except tuna within the U.S. fishery conservation zone (FCZ)" (U.S. Dept of Commerce [NOAA] 1982:1). The states retain jurisdiction over the first 3 miles offshore, the territorial sea, with the federal government having jurisdiction over the remaining 197 miles.

The act also set up eight Regional Fishery Management Councils. Conservation demanded the regulation of both foreign and domestic fishing. The eight councils were to develop and implement management plans for the various species within their management areas. Pacific Northwest fishing is under the jurisdiction of two councils: the Pacific Fishery Management Council, with representatives from the states of California, Oregon, Washington, and Idaho; and the North Pacific Fishery Management Council, with representatives from Alaska, Washington, and Oregon. Each council is responsible for the fishery resources within the Fisheries Conservation Zone (FCZ) adjacent to its member states.

The Pacific Fishery Management Council's initial salmon management plan, implemented in 1977, affected United States troll and charter fishermen as well as Canadians operating in United States waters. The plan was amended yearly until 1981, when it was modified substantially. The 1979 and 1980 regulations had specified merely that chinook could be taken between May 1 and May 31, and that all species of salmon could be taken between July 1 and September 8 (1979), or July 15 to September 8 (1980), with "emergency closures" and "adjustments" taking place from time to time. The 1981 restrictions, which were implemented to protect coastal and weaker salmon runs, established specific fishing seasons, areas, and harvest quotas for ocean salmon fishing. These restrictions were very unpopular with nontreaty fishermen. Similarly, the North Pacific Fishery Management Council implemented its salmon plan in the fall of 1979, and it imposed emergency regulations on high-seas fishing off the coast of Alaska for all salmon except coho in June 1981. These regulations shortened the fishing seasons and lowered the number of salmon that could be taken.

Both councils have also provided plans for the management of species other than salmon. The North Pacific Fishery Management Council implemented its initial tanner-crab management plan in May 1978. The plan was amended several times and was revised once more by 1981. The year 1981 also saw the development of a Bering Sea/Aleutian Islands king-crab management plan. Both councils developed plans regarding groundfish. In addition, the North Pacific Fishery Management Council developed a herring plan, and the Pacific Fishery Management Council, an anchovy plan.

The Magnuson Fishery Conservation and Management Act of 1976 had a great impact on Pacific Northwest fishing. While most fishermen were in favor of excluding foreign fishing from U.S. waters, they were not prepared for federal management of their own activities. Compounding this, the act was implemented just a few years after several court decisions which had also profoundly affected Pacific Northwest fishing. These court decisions all dealt with Indian treaty fishing rights. The most important case, *United States* v. *Washington*, had been decided February 12, 1974. It is commonly known as "the Boldt Decision."

The Boldt Decision provides that those Indian tribes party to a series of treaties signed in 1854 and 1855 are entitled, in the words of the Medicine Creek Treaty, to "the right of taking fish at all usual and accustomed grounds and stations . . . in common with all citizens of the Territory." "In common with" is interpreted to mean that treaty Indians are entitled to take 50 percent of all harvestable fish. The Boldt Decision further provides that the state has no regulatory jurisdiction over off-reservation fishing by Indians, except for conservation; that the state cannot discriminate against Indians in its laws and practices; that the state must conform to "the requirements of due process" in regulating Indian fishing; and that the treaty tribes "may regulate their own members' off-reservation fishing and thereby avoid almost all state regulations if they meet certain qualifications and conditions" (American Friends Service Committee 1970:xiv). In response to these conditions, in 1974 the case area treaty tribes set up the Northwest Indian Fisheries Commission as a coordinating body for five newly formed treaty councils.[7] Each council is composed of those tribes included under a single treaty. Representatives from each council sit on the commission. Commission procedures and Indian efforts in self-regulation are illustrated below in the life history of Linda Jones.

The years immediately following the Boldt Decision were ones of turmoil. Nontreaty fishermen organized demonstrations, much as American Indians had during the 1960s and early 1970s, and some engaged in illegal fishing in protest during the years of 1975 through 1978. The decision was appealed to the U.S. Ninth Circuit Court of Appeals in April 1974, and was upheld by that court on June 4, 1975. After first declining to review the case in 1976, the U.S. Supreme Court accepted it for review in 1978. On July 2, 1979, the U.S. Supreme Court upheld the Boldt Decision and the 50 percent allocation of harvestable fish, with the proviso that fish caught on reservations and those caught for ceremonial or subsistence purposes were to be included in the treaty share.

During the years between the initial Boldt Decision (1974) and the U.S. Supreme Court's affirmation (1979), the state of Washington was either unwilling or unable to enforce the provisions of the decision. In particular, the

7. A comparable decision (1969) affecting Columbia River fishing is known as the Belloni Decision (*Sohappy* v. *Smith*, Civil no. 68–409, consolidated with *United States* v. *Oregon*, Civil no. 68–513). The tribes party to this decision formed the Columbia River Inter-Tribal Fish Commission in 1977.

Washington Supreme Court ruled that the Department of Fisheries did not have the authority to allocate fish between treaty and nontreaty fishermen. The Department of Fisheries reverted to pre-1974 fishery regulations. Consequently, in 1977 the federal court assumed management and regulatory responsibility, first over treaty fishing, then over nontreaty fishing in western Washington. A footnote to the U.S. Supreme Court's Majority Decision states: "the state's extraordinary machinations in resisting the [1974] decree have forced the district court to take over a large share of the management of the state's fishery in order to enforce its decrees" (full text, *Northwest Indian News*, 1979:18 [Note 36]). The majority decision further states that: "the District Court has the power to undertake the necessary remedial steps and to enlist the aid of the appropriate federal-law enforcement agents in carrying out those steps" (Section VII:17).

What is known as Phase II of the Boldt Decision was handed down September 26, 1980, and involved issues that were included in the 1974 case but were reserved for later consideration. The Phase II decision provided that: treaty Indians are entitled to 50 percent of state hatchery-bred and artificially propagated salmon and steelhead, as these fish are meant to replace fish lost to natural runs; and that they have the right to protect the environment surrounding their fishing waters, as their treaty rights would be meaningless if there were no fish to take. In the most extreme interpretation of this provision for environmental protection, treaty Indians could veto Department of Fisheries policies as well as any development projects that might threaten their fisheries environments.

Washington State requested a review of the Phase II decision by the U.S. Ninth Circuit Court of Appeals. In its April 29, 1985, decision, the circuit court upheld the ruling that treaty fishermen are entitled to a share of hatchery-bred and artificially propagated fish. The state requested that the U.S. Supreme Court review this decision but the Supreme Court declined. Although the hatchery issue is now settled, litigation continues on the environmental issue. The U.S. Ninth Circuit Court of Appeals vacated the district court's decision and remanded it to that court for further proceedings. It did so on the basis that the state's precise obligations and duties with regard to the fisheries environment "would depend for their definition and articulation upon concrete facts which underlie a dispute in a particular case" (U.S. Ninth Circuit Court of Appeals 1985:1354). At this writing, the environmental issue is under a stay of proceedings in the U.S. District Court of Western Washington and will come up for consideration in mid-1988.

The effects of the Boldt Decision have been far-reaching. Although the case initially involved only salmon and steelhead, the court indicated that Indian treaty rights are not limited to these species alone. In 1975 Judge Boldt expanded the *United States* v. *Washington* decision to include herring. In addition, the U.S. Supreme Court Majority Decision states that:

any fish (1) taken in Washington waters or in the United States waters off the coast of Washington and (2) taken from runs of fish that pass through the Indians' usual and accustomed fishing grounds and (3) taken by either members of the Indian tribes that are parties to this litigation, on the one hand, or by non-Indian citizens of Washington, on the other hand, shall count against that party's respective share of the fish [Section V:15].

Given salmon migration patterns, this could mean that fish caught by Washington State residents in Alaskan waters could be counted as part of the nontreaty fishermen's 50 percent allocation. The difficulty lies in determining which Alaskan-caught fish derive from runs passing through treaty Indians' usual and accustomed fishing grounds.

The principles delineated by the initial Boldt Decision and the U.S. Supreme Court Majority Decision upholding it also affect fishermen through the interaction of these decisions with other legislation. The Pacific Fishery Management Council, created by the Magnuson Fishery Conservation and Management Act of 1976, sets ocean fishery regulations. Treaty fishermen challenged council regulations in 1979 and again in 1981 on the grounds that ocean regulations should protect weaker runs, even if this should mean that fishing on stronger runs be curtailed. Treaty fishermen are predominantly river fishermen, and their share comes solely from those runs returning to their rivers. Any overfishing on their stocks therefore denies them their allocation. This stance particularly affects Washington State trollers. Trolling is a hook-and-line ocean fishery, and trollers fish on mixed stocks of salmon—that is, salmon that have not yet broken up into runs to return to their fresh-water places of origin. Council regulations for the 1981 season reflected the treaty tribes' point of view.

The implementation of the Boldt Decision has been very hard on nontreaty commercial fishermen, as is amply demonstrated in the life histories in this volume. The amount of time they can fish has been drastically reduced. Because most treaty tribes are river fishermen, while most nontreaty fishermen fish on Puget Sound, San Juan Island waters, or the ocean, and because of the smaller number of salmon available overall, marine commercial fishermen have had to take fewer fish to allow sufficient escapement for sportsmen, conservation, and treaty tribes' shares. The nontreaty commercial fishermen's harvest and income therefore have dropped, and many have not been able to continue fishing. According to a recent report, the number of nontreaty salmon vessels licensed to fish in state waters has declined by 49 percent since 1975 (Natural Resources Consultants 1986:22).

Because of the Boldt Decision, the state also has had to alter its record-keeping system. Early in the development of the fishing industry, the state had kept track of the numbers or pounds of fish landed as an aid in tax collection. In 1951 a "fish-ticket system" for counting fish was implemented.

Generally, when fish or shellfish of any species are delivered to a buyer, the buyer fills out a fish-receiving ticket. The ticket lists the number of fish and pounds for some species (including salmon), and only the number of pounds for other species. After both the buyer and the fisherman sign the ticket, the buyer keeps a copy, gives a copy to the fishermen, and sends a copy to the state. All tickets are submitted on a daily basis. Since the Boldt Decision, the state has had to keep track of individual fishermen as well as fish. The buyer indicates on the ticket whether the fisherman is Indian or non-Indian, treaty or nontreaty, and whether the catch was made on-reservation or off-reservation. The appropriate treaty tribe also receives a copy of the ticket. The state processes all this information, keeping track of individual fishermen and the category to which each belongs in order to allocate fish in accordance with the Boldt Decision and to determine just how many fish can be taken while still preserving the resource. The current system began in 1976.

In 1974, at a time when the number of salmon was declining, the Boldt Decision shifted a larger share of the fishery to treaty Indians. It became immediately apparent that the level of fishing effort must be decreased. The state of Washington placed a moratorium on the issuing of new commercial salmon-vessel fishing licenses in 1974 and expanded it to include charter-boat licenses in 1977.[8] The 1974 moratorium put a ceiling on the number of vessels able to fish commercially, but it did nothing to reduce these numbers, so the state instituted a buy-back program in 1975 using federal monies. Under the initial program, the state bought both fishing licenses and vessels in order to remove them from the fishery, and participation was restricted to net-fishermen directly affected by the Boldt Decision. The current program, implemented in 1980, buys only fishing licenses. The vessels themselves are retained by their owners or are sold with the stipulation that they cannot be used in the state's salmon fisheries for ten years. In addition, participation was expanded in 1980 to include all state commercial (including charter) fishermen, regardless of gear type or geographic area of concentration.

The situation is similar in the states of Oregon and Alaska. Oregon imposed a moratorium on new commercial salmon-vessel permits in 1979 and also has buy-back programs. In Alaska, the limited-entry permit system was passed by the state legislature in 1973. Salmon fisheries, particularly those in Southeast Alaska, were the first to come under the system. Fishermen were issued "interim-use" permits (based on past participation in the fisheries) until the details of awarding the permanent permits could be worked out. Permanent salmon permits were issued in 1975.

---

8. In the state of Washington, as in Oregon, a fishing license is issued to a vessel. In Alaska, the fishing license, or permit, is issued only to "natural" persons and not to vessels or corporations.

Thus, in all three states no new licenses or permits are available from the state governments. Those wishing to enter the fishery for the first time must buy their licenses and permits from existing holders, usually for extremely high prices.

Because of the Magnuson Fishery Conservation and Management Act of 1976, management plans for other species besides salmon are under development. The extreme pressure on the salmon resource dictated that that issue be addressed first. In addition, any limited-entry programs in place for salmon fishermen may eventually be applied to other species as well. This has already happened in Alaska, where the limited-entry plan was applied to various roe herring fisheries in 1977, 1978, and 1981, and to Southeast Alaska king and tanner crab fisheries in 1984. King and tanner crab fisheries occurring farther offshore come under federal jurisdiction.

The fisheries in all coastal states are now managed jointly by the regional councils set up by the 1976 act and by the states themselves, in accordance with any international treaties, such as the one between Canada and the United States, and treaties with Indians. The states have jurisdiction over the territorial sea, while the councils have jurisdiction beyond the three miles. All agencies are supposed to assume responsibility for implementing the councils' fishery management plans. Disagreements between agencies can occur, however, as in 1981, when California opened its coastal waters to trollers while the larger Fisheries Conservation Zone (FCZ) was closed. All these agencies and plans to regulate fishing would be unnecessary if the supply of fish were limitless, as was once thought to be the case.

Choices

The women whose lives are told in this book have inherited the outcomes of the historical processes just described. They are also critical actors in the current arenas of change. Fishermen can no longer simply cast off their boats at the most desirable times to fish in the most desirable locales; they are not engaged in fully free enterprise as stereotypical daring and adventurous entrepreneurs. They are bound by regulations that control where, when, and with what gear they may fish. Consequently, being a successful fisherman means that one must be aware of the frequently shifting rules of the game. When the rules interfere with making a living at fishing, one's choices are few: to abandon fishing; to become involved in political activities designed to change the rules; or to adjust to the shifting rules as expeditiously as possible. Only one of the women in this book (Christina Jefferson) has abandoned the industry, and she has remained in a marine-related occupation. Some have become involved in political activities in an effort to bring about changes; others have adjusted their lives many times in order to stay in the fishing industry. Linda Jones continues to work for the Tulalip Tribes, but in a different administrative capacity.

The Time Line that follows outlines historical aspects of the Pacific Northwest commercial fishing industry and locates the time at which each of the ten women entered the industry.

Time Line

1823   Hudson's Bay Company packs salted salmon on Columbia River.

1830   Russians pack salted Alaska salmon for export.

1833   Hudson's Bay Company sets up Nisqually post in southern Puget Sound.

1840s   Indian fishing increases to meet demand.

1853   First commercial fish operation by non-Indians at Seattle, "Doc" Maynard sends salted salmon to San Francisco.

1854–55   Medicine Creek and other treaties signed, promising treaty Indians the right of taking fish at their "usual and accustomed grounds and stations."

1864   First West Coast salmon cannery, in Sacramento, California.

1867   Alaska purchased from Russia.

1877   First Puget Sound salmon cannery, in Mukilteo, Washington.

1878   First two Alaskan canneries, in Southeast Alaska.

1880   Beginning of Puget Sound settlers' use of traps as primary fishing method.

1884   Successful shipping of fresh salmon in refrigerated rail cars from Oregon to East Coast.

1889   Washington becomes a state.

1890s   Introduction of steam-driven boats in fisheries.

1900s   Indians lose major fishing locations to settlers and to state regulations.

1903   Salmon-cleaning machine developed in Bellingham, Washington.

1905   Refrigerated trucks first used.

1906   Albacore first canned in California.

1912   Federal government begins management of Alaskan salmon.

1913   Rock slide at Hell's Gate Canyon on Fraser River.

1914   Purse seining becomes economically dominant type of fishing in Puget Sound.

1915–16   Gasoline-powered vessels first used.

1920s   Trollers in California begin fishing albacore.

1921   Washington State Department of Fisheries established.

1924   *Gladys Olsen* first works in a clam cannery.

1925   Steelhead declared a game fish in fresh water in Washington.
        Mechanical-refrigeration truck first used.
1926   Albacore vanish due to warm water temperatures, probably due
        to an *El Niño* episode.
1930s   United States begins commercial king crabbing in Alaska.
        Japanese begin king crabbing and trawling for bottomfish near
        Aleutian Islands.
1933   Passage of Initiative 62 establishes Washington State Depart-
        ment of Game.
1934   Passage of Initiative 77 abolishes use of salmon traps in Wash-
        ington.
1935   Albacore trolling begins in Pacific Northwest.
1937   International Pacific Salmon Fisheries Commission established.
1941   United States enters World War II.
1942   *Tink Mosness's* husband begins to fish commercially.
        Gladys Olsen's husband begins to fish commercially.
        Japanese bomb Dutch Harbor in June.
1943   Battle to reclaim Attu from Japanese.
        Loran-A becomes fully operational.
1945   World War II ends
        Regulation of Fraser River sockeye begins.
        Radar first available commercially.
1948   *Mars Jones* begins to fish commercially.
        Mechanical tuna-canning developed.
1950   Development of commercial Alaskan king crabbing.
        Introduction of seine drum.
1951   Introduction of fish-receiving ticket system in Washington.
1952   *Marya Moses* begins to fish commercially.
        Japanese begin first high-seas salmon fishery near Aleutian Is-
        lands.
        International North Pacific Fisheries Convention treaty is signed
        between United States, Canada, and Japan.
1954   *Tink Mosness* becomes active in fisheries politics.
        Introduction of nylon fishing nets.
1955   *Lois Engelson* begins to fish commercially.
        First use of Puretic block on seiners.
1956   *Helen Gau* begins to fish commercially.
1957   International Pacific Salmon Fisheries Commission jurisdiction
        extends to pink salmon species and to Puget Sound area.
1958   *El Niño* drives Fraser River salmon off usual migration route.
1959   Alaska becomes a state.
        Russians begin to fish near U.S. Pacific Coast.
        El Niño causes albacore to go as far north as Gulf of Alaska.

1964   Earthquake and tsunami.

Canada implements "twelve-mile limit."

1966   United States adds nine-mile contiguous zone to existing three-mile territorial sea, resulting in twelve-mile limit.

1967   Development of Alaskan commercial tanner-crab fishery and processing techniques.

1968   First Alaska State closure of king crab season.

1969   South Koreans begin gillnetting for salmon of U.S. origin.

1970   Development of vessels specifically for king crabbing.

Canada and United States implement reciprocal fisheries agreements periodically until 1978.

1972   *Evie Hansen* begins to fish commercially.

Christina Jefferson begins fish processing.

1973   Alaskan limited-entry legislation enacted.

Marti Castle enters fish-processing business.

1974   Boldt Decision (*United States* v. *Washington*).

Evie Hansen becomes active in fisheries politics.

Washington State moratorium on commercial salmon licenses.

Boldt Decision appealed by Washington State.

Northwest Indian Fisheries Commission established.

1975   Boldt Decision expanded to include herring.

Boldt Decision upheld in U.S. Ninth Circuit Court of Appeals.

Funds allocated by Washington State Legislature for state buy-back program.

U.S. Coast Guard begins replacing Loran-A with Loran-C.

1976   Beginning of current Washington State fish-receiving ticket system.

Magnuson Fishery Conservation and Management Act of 1976 establishes 200-mile Fisheries Conservation Zone and eight regional management councils.

Linda Jones begins work for Northwest Indian Fisheries Commission.

1976–77   *El Niño* episode.

1976–78   Illegal fishing by nontreaty fishermen as protest against Boldt Decision.

1977   Washington State moratorium on issuing new charter boat licenses.

Federal court assumes management of Washington treaty fisheries.

1981   Implementation of Pacific Council salmon management plan.

1978   Implementation of North Pacific Council tanner crab management plan.

U.S. Supreme Court agrees to review Boldt Decision.

1979   U.S. Supreme Court upholds Boldt Decision.

            California and Oregon begin moratorium on new commercial
                salmon-vessel permits.

            Implementation of North Pacific Council salmon management
                plan.

            El Niño episode.

1980   Boldt Decision, Phase II (*United States* v. *Washington*).

1981   Pacific Council salmon management plan substantially modi-
                fied.

1981   Implementation of Bering Sea/Aleutian Islands king-crab man-
                agement plan.

1982–83   *El Niño* drives Fraser River sockeye salmon off usual migration
                route.

1984   Fishermen apply for Small Business Administration disaster loans
                due to effects of *El Niño*.

1985   New United States-Canada salmon treaty concluded.

            International Pacific Salmon Fisheries Commission absorbed by
                new United States-Canada Salmon Commission.

The Life Histories

Tulalip fishing fleet (*photo by Sue-Ellen Jacobs*)

Mars Jones

Mars Jones

Mars Jones lived on a farm in eastern Washington until first grade. She spent the remainder of her childhood in West Seattle. After finishing high school, Mars married her husband Gaylord in Seattle in 1937. They soon started a family. After Gaylord's discharge from the army, he and Mars made their living as jazz musicians. Gaylord played piano, and Mars played percussion and sang with the band. Gaylord loved sportfishing, and Mars came to love it, too. So they soon decided to try commercial fishing.

Beginning Fishing

[My husband] has been a sportfisherman . . . from the time he was little. His father died and his uncles took him fishing. He trout fished originally. Then he went to salmon fishing, and I got in on that. To me, salmon fishing was more exciting than trout fishing, because one fish filled up the dinner table. [Gaylord] was in the army for a couple of years. He was a piano player in the USO units at Fort Lewis. [We] sportfished every [weekend] you could get off. When he got out of the service . . . we got to thinking about fishing . . . all the days we didn't work nights. Pretty soon we decided it would be more profitable to fish than it would to be in the music business in the summer. And it was more fun. We got tired of the smoky night clubs and everything. In 1948 [we started to fish commercially].

We got a small boat and we were a kelper. A kelper . . . fishes along the beaches. He can't go out to sea very far, so he has to fish along the beach; he's a day fisherman. He sells his fish every day, and he can't stay overnight [because the boat isn't big enough and doesn't have] ice or refrigeration. We used to be able to mooch and catch them with sports gear, but they changed the laws and we had to go to commercial gear to catch commercial fish. A kelper . . . handlines [with] a pole . . . like a sportfisherman, and he could fish any way in those days, but now you have to have fixed gear even if you have a small boat.

Entering the Charter Boat Business

Mars and her husband began chartering after they bought a larger boat in 1952. Gaylord ran the boat with Mars's assistance. Later Mars began chartering on her own.

We could see where you had to have a bigger boat [to make it financially worthwhile]. We used to fill our little boat, and we didn't want to overload it. [During] the time you go in to sell them because you're afraid to take any more fish—and the fish are biting—you lose out. So you had to have a bigger boat and better equipment. There was no bathroom on those little boats [and] no place to cook. [Our season was between] when school let out and when school started up [June through August]—a very short season—and you don't really make enough money that way. You were better off to start in those days . . . March 1 [and go] to October 31. We could never do that until our family grew up. We were musicians for the rest of the year then.

When we first started, we charged our fuel bill. We learned not to do that. That's the biggest mistake we made the first year. It was hard to pay that fuel bill during the winter plus all our other bills. You pay it right as you go along, and the same way with other expenses when you're out there. Pay as you go along; don't charge it. That was probably the best lesson we learned.

We got a bigger boat in 1952. We started chartering just one boat. I helped him [Gaylord] for [three] years. In 1955 then I started running charters by myself around Neah Bay. It's probably the most beautiful spot in the world to fish. The fish were big, and you didn't have to go over a bar. You're six miles from the ocean. I got my Coast Guard license and everything.

Some guy at the Neah Bay talked me into running his boat. I said, "Well, I know I *could.*" But I thought it was a big risk for him, because all he knew was what he saw I did with my husband. There were times when my husband got sick . . . so I'd run the trip, but he'd stay on board . . . in the bunk. And then this guy wanted me to run his boat for him, and I said, "I can't without a license, somebody has to be on board with a license." So he says, "Well, go get it." I said, "Well, they won't give it to a woman." He says, "Well, you don't know if you don't try." So I went down and went right through it.

In those days [the test] was oral. The fellow in charge was quite a tough guy. He used to send the guys home two or three times to study more because they didn't give the right answers. I had heard the guys talking about their license. I memorize things easily. Sometimes things [didn't] make sense, but I'd memorize it because when [Gaylord] took his test, I'm the one that asked him the questions. I didn't even have to look up the answers because I'd memorized them. [When] I went down and took my license [the fellow in charge] said I was the first woman to try. He said, "What do you want it for?" I said [that] sometimes my husband gets a migraine, which was the truth, and [that] I didn't like to turn people down. I wanted to be sure that I was doing the right thing by running the trips, and I wanted my license. He said, "Sounds sensible enough." So I got my license. Every five years you get it renewed; they want to be sure you know the [latest rules and regulations]. The Coast Guards never charge a fee [for the license]; why, I don't know. There should be a fee, because it's a lot of hassle for them. But, of course, we're socked enough for fees for every time we turn around anyway.

Then I got this other fellow wanted me to run his boat. So I started running boats, and I had three or four boats that I ran. I had to be able to jump on board a different vessel. Sometimes it was a diesel, sometimes it was a gas engine, and I never took one unless I looked the engine over. A mechanic I'm not, but I learned [about boat engines]. Being a musician, I could always hear if it didn't sound right. So if it wasn't firing right, I got so I could adjust carburetors and things like that. Everybody [has the engine malfunctioning some time]. I've probably been towed in three times. You have more breakdowns with a gas engine than you ever will with a diesel. I blew a rod one time in a gas engine . . . and I got towed in. But in the meantime, while we were waiting for our tow, we finished getting our fish. We just drifted and mooched—got our fish. I never had problems with a diesel—knock on wood. In those days we didn't have radar. . . . We didn't have a radio. All we had was our compass and a leadline. And what fog we had! Maybe three-quarters of the time during August. And towards the last there, without the radar, I was running three trips a day and never seen the land. But you learn the water best that way. All radar does is verify the fact that you know where you are. [Then there's the knowledge of where to find the fish]; if you're out there every day you keep on top of them. After so many years of fishing out there, you have an instinct. If they aren't there one day where you're looking for them, you know they've got to be in another corner. So then you go look in another corner the next day.

When I started, back in '55, people would come to the dock—they'd see it was a woman, and it'd be foggy. I had four guys one time—they looked at me, and they looked at the fog, and they said, "I'm not going out with a woman in the fog." I said, "Suits me fine, because I'm going fishing." And I cast off from the dock and left them standing there, and I made *way* more money going out and commercial fishing—I had a commercial license—than I would have if I took them out. But I knew that I was building a job for the future. I knew—instinct, I guess—that someday commercial fishing would peter out. And that's what's happening.

Then I decided I'd rather have my own boat. That was about 1958 I got my first boat. I had that one for about fifteen years. And then I got the *Satin Doll I,* and then [my current one]. The only thing [my husband] did on my boat was to keep the engine running. I did all the sanding and painting and bottom-painting—all the maintenance. Then when I had number one built . . . the bank didn't want to finance it in my name. We had a little trouble over that, and I told them I would go where I could find a bank that would finance it in my name. I said, "Even though my husband backs me up, I want it in my name." So we got it in my name.

*Mars and her husband moved to Port Angeles in 1962; they retired from profes-
sional music and became year-round fishermen. Mars takes charter boats out of
Port Angeles in the winter. In the summer she charters out of Neah Bay and lives
on her boat. Gaylord takes charter boats in the winter and trolls offshore from
Neah Bay in the summer. Every seven or eight days he comes in to sell his catch
and take on supplies. Mars and he communicate regularly by radio.*

*By mixing chartering and commercial fishing, Mars and her husband are able
to insure themselves against the ups and downs of the economy in either enterprise.
When Mars first took out charter boats she also had a commercial fishing license.
At that time she could make more money fishing commercially than chartering,
but she saw more of a future in chartering. The income from part-time commercial
fishing was essential for making her boat payments before she had built up a regular
charter clientele.*

*The Hood Canal Bridge connecting the Seattle area to the Olympic Peninsula
was blown down during a storm in 1979. During the years it was down, access*

to the Olympic Peninsula was limited. This damaged the charter boat industry. Mars got a commercial bottomfish license in order to maintain her income until the sportfishermen could more easily reach Port Angeles for charter boat fishing.

Running a Charter Business

Mars prefers chartering for sportfishing, while her husband prefers trolling. Mars found that she missed all the people when she tried full-time commercial fishing one season. Mars has an automatic pilot on her boat, and in the winter, with fewer clientele, she works alone. In the summer she usually takes on a deckhand. The deckhands have included her granddaughter and a pre-med student. Of the student, Mars says, "She's a one-man army! She should have her license, in fact." Mars enthusiastically describes what is involved in running her charter fishing business.

We get an awful lot of repeat business, and they get very upset if they don't get on *your* boat.[1] But the people come aboard . . . and I usually tell them to bring their lunches and whatever they want to drink. Everything else is furnished—their tackle, their gear, their bait. During the winter we furnish the fish cleaning, the fish bags. In the summer they have to pay for fish cleaning . . . unless they want to do it themselves . . . and bring their own bags. I see that they have punch cards, and they have to buy salmon licenses. [In the winter] I have to furnish these things; in the summer the [Big Salmon] Resort does that in Neah Bay. It's owned and operated by an Indian and a white man. The punch card has a Washington State license on it. I explain to people how to do it. Most of them know by now, unless you get an out-of-stater. I had a boy from England the other day, and I had to help him with it a little bit. Your punch card is free, but your stamps are $1.25 a day, or $3.25 for all year. You get thirty punches; you punch for every legal fish you catch. Right now [in February] the silvers are too small [to be legal], so what we're catching are all blackmouth—that's the immature king salmon. The other day we had a freak thing happen. We had a capon. This was a hooknose silver. It should've been a male, and there was no indication that he was a male in his entrails, so he was a capon, I guess. It's the only thing I can think of calling him. He was an "it," and he was nine pounds.

I net the fish, and bait the hooks, and keep the gear going, and explain how it works—keep everybody happy. If somebody's catching all the fish, I make him quit when he's got three. Some of the places don't do that, but

---

1. In the winter, sportfishermen can contact Mars and her husband directly; in the summer, reservations must be made through Big Salmon Resort.

we figure everybody pays their own way and they should have a chance to catch their own fish, especially on mixed parties [with] individual people. There's bound to be a hot pole—some days they'll all work, and other days there'll be one pole catching most of the fish—so you've got to rotate. Some people are hard to get fish for, and other people no matter what pole, what position they take on the boat, or what kind of a day it is, they'll get a limit. It must be . . . positive thinking or something. We get about 98 percent limits here in Port Angeles and . . . we used to get almost 100 percent at Neah Bay. But over the years the fishing pressure's getting heavier and heavier . . . and therefore you get about 85 percent limits, and that's a pretty good guess. [Up at Neah Bay] in the summer you get into your big schools of fish, and I have [had] fifteen people [catch their limit] in an hour and a half. And many trips I had the three fish limit by two silvers and one humpie, or two kings and one humpie. Pinks, they call them, and they're good fish.

Over the years I think I'm getting more women now, and sometimes I have an all-hen party. I used to get a big group from Boeing—take the whole boat. They were a lot of fun. [*Could they fish?–ED.*] Oh, you bet! I'd say it's how many times a person's gone fishing, how they handle their fish. Some men are just . . . clumsier with the gear than some women. Some women, no matter what pole you give them . . . they just know how to handle it. It's just . . . up to the individual. Then too . . . no matter what kind of business you're in, being a woman, you've got to handle men with kid gloves. You cannot squelch their male egos. If they've got their heads set on something, let them do it. [Like] some guy—you tell him how much line he should be putting out—he isn't going to listen. You have to let him get hung on the bottom and lose gear and let him find out. Some of the guys that you really think you're going to have trouble with—if you kid them along a little, and then you begin to feel them out a little, you get to know them and they turn out pretty nice. My hair [used to be] a little redder than it is now. It used to bug me, but I learned to swallow it. . . . . I had to learn these things, and I tried to teach my daughters how to cope with their husbands. I think it has helped.

[Sometimes] they drop things overboard and break things, and I just figure it's one more thing I'll deduct from my income tax. I haven't [charged them] because lots of times . . . it looks like it might have been accidental, and I figure it's one of those things. Like you'll get an old man—he'll pick up a pole, and his hands may be stiff or something. He'll drop it, and just the whole thing goes over. What can you do? The poor old guy probably hasn't got enough money to ———. I have $100 in every one of those poles. Sometimes I spend hours repairing the gear for the next day.

[Other charter boat owners charge for lost equipment if] they get somebody that's drunk. I won't take them. The Coast Guard told me many years ago, "Refuse to take anybody that's already drunk; it will save you a lot of trouble. If you ever can't handle those guys out there, just call your local Coast Guard." They'll come and remove them . . . [because] it's a public conveyance. I don't think I get one drunk in two or three years. For one thing they can't drink too much or they get sick if it's rough. But I think . . . what's happening to America is that they're not drinking water—they're drinking beer. I've never seen so much beer consumed in all my life. Some of them will bring . . . on board a huge deal, and it'll be just loaded with beer! And where are their sandwiches? They don't bring sandwiches. I don't understand it. Now, the ones that bring food are usually an easier group to get along with.

Now, speaking of getting fish for the thrill of it, August becomes a meat month. People are after meat . . . not the sport. And that's one month of the whole year I would like to cancel out and take my vacation. People are just fish hogs. They want the biggest and the most. They come aboard . . . in droves. You have to turn down people every day . . . that don't have a reservation. The ones that do get aboard, they get grabby. They'll swipe other people's fish; they'll swipe your gear; they'll take off your lures . . . they take stuff out of your drawers. Talk about "sportsmen fish"—sportsmen!? I don't know. They want three twenty-pounders or above, and you bring in this twenty-inch fish and they want to throw it back. What are you going to do? Maybe that's all you have to offer them that day. Sportsmanship is the most important thing in going fishing. . . . If you're going to hook a twenty-incher you're supposed to keep it. And sometimes I let them catch my limit if the fishing's good, and then I'll swap them for the little ones. If I ever had to hire someone for an important job, I'd take them fishing first—they really bring out their personality. But the rest of the year I love the people. They have so many different backgrounds, and I always ask them about themselves and where they're from. And I get some people from the Orient that can't speak English, but *they* know how to fish. No matter what kind of fishing pole you give them, they know what to do.

Some aspects of Mars's charter fishing business are regulated by state and federal agencies and by charter fishing organizations.

In the summer I've gotten up to fifteen [charter passengers on at one time]. [My boat] is a twenty-two passenger. The fisheries department gave us a limitation, and I can only operate sixteen. Too many people [were] catching fish, so each boat was cut back [as to] how many they can take. They gave me sixteen . . . unless I have a jigging group. They jig for bottomfish. They bring their own gear . . . and I can take up to twenty-two. I can also take twenty-two sightseers, which I do once in a while. Cape Flattery is very scenic, and I can spend two hours out there just showing them the caves and islands and rocks. It's a photographer's delight and a bird-watcher's delight to get out there and see these things. Twelve people at one time is [about average] in the summer.

The Charter Boat Association [meeting] was in January, and for the winter rates it's gone up to $38 plus tax per person. In the summer it's going to be a little more. What they've set I'm not quite sure, but I think it's going to be close to $45. It hasn't been completely set; they've got to get together with the other resorts. It's the winter rates for people over seventy who do not, by the way, have to buy a license. We can't price our older senior citizens right out of fishing; we've got to help them a little. I'd say in the wintertime we probably get 80 percent older people . . . mostly because Sequim is so close, and that's a good retirement center. And we have an unusual thing here; if it's blowing southwest or southeast, we don't get the wind. We have the [Olympic] Mountains to protect us.

I'm registered for twenty miles offshore, any direction—twenty miles from a safe harbor—and that includes your six miles in the Straits [of Juan de Fuca]. That's far enough to go. Any farther . . . is too far to go when you've got all those people on board. My ETA [estimated time of arrival] at the dock is usually two-thirty, three o'clock. This year they voted to get them in about one, one-thirty in the afternoon. I can't do much about that; they all voted on that. [We leave at] daybreak, and that's early in June; in August you get a little more sleep. My average winter trip—we're back before noon. But there are some days when the fish are slow and the sun comes out and it's beautiful—and then we try to be back at three. In the winter we book our own [people]. It's repeat business. We've been in business many, many years, and we do a lot of advertising, too—in *Outdoor Press* (that's over in the Spokane area), *Fishing and Hunting News*, then we have a Jimmy-come-lately from Sequim, and Seattle [telephone] directory, I believe, has Big Salmon [Resort] in it.

[I belong to the Washington State Commercial Passenger Fishing Vessel Association]. We originated the Charter Boat Association here [in Port Angeles], our group did. And then we had a fellow here was a real sharp guy, and he persuaded Westport and Ilwaco to organize. We've been organized for quite a while, probably '72, '73. [It's for] charter boat operators. [You don't have to be an owner.] Resorts buy a membership. One of the shipyards bought in a membership. That way they get in on all the dope on what's happening in the Charter Boat Association. We retain a lawyer to watch the lobbyists in Olympia.

Our branch meets once a year, but we have meetings that we can go to at Ilwaco and Westport. They have quite a few meetings a year. And then we send representatives to the National Fishermen's meetings and all those.

We try to keep good feelings between all the groups, because we think our main thing is to stick together, regardless of what kind of fishermen. Each group should know what the other one wants, and they should stand together. You need everybody together to get what we need and what we expect of a season. I was secretary-treasurer, and I went to the meetings, but we're so busy on this winter meeting stuff that ———— . It was my husband's idea to run for that humpie thing—[fish] till we get three fish—two silvers and one humpie, or two kings, or one king and a silver and a humpie. The president of our local here on the [Olympic] Peninsula, he went after it and he did a good job of it. And then we know what the trollers have done, because [my husband] belongs to the Trollers Association. We read a lot too [about] what's going on. We have *Fish Boat*, and there's a couple other magazines that they send us. And then we find out what the Canadians are doing, and they find out what we're doing.

Relations with Commercial Fishermen

We've had no trouble [with commercial fishermen], but I did have a fellow try to ram me last fall. They closed the ocean, and I came in from Neah Bay and I ran a trip. It was my first trip out, and I couldn't locate the fish. I went all over, and everybody said, "Well, there's some silvers out in the middle of the Straits—if you're lucky you'll find them." I went all over. It was noon, and I didn't have a fish in the boat. I moved in where we fished for them along the shore in the winter. I told my husband [on the radio], "I'm going in there and I don't care—I've got to get something in this boat." And he says, "Well, let me know." I moved in there and I started catching fish like mad. And he was coming—he pulled his gear and started to come up there.

In the meantime, I saw these two boats running in from the ocean, because it was closed to commercial fishing and closed also to sportfishing. The next thing I knew, this guy had departed from his line of course and was headed for midships on my boat. I kept watching him. I was netting fish, and baiting them up, and throwing them out—and he kept coming and coming. And there was a little boat alongside of me [with] just barely enough room for him to come between us. I thought, "Well, I wonder what he's up to." I got to thinking maybe he's a hothead. So I watched it. I ran up and I only had one engine on, and sure enough he was going to try to ram my boat. I didn't have time to start the other engine, so I just revved up and got out of the way, and he ran across the back end of my lines. He would've cut all my lines off.

But when I got in, I got his number and I tried to report it to the Coast Guard. Now, this is what's so goofy about the Coast Guard. It takes ten sheets of paper to fill out a report, and then you've got to take it to court. I didn't take it to court because I figured with all the red tape it wasn't worth it. But I stopped and talked to the fellow, and I said, "What's the big idea trying to ram me?" And he says, "Fishing season's closed." I said, "It doesn't affect me." And he wouldn't talk to me any more. It was open inside the ocean.

The inside of the ocean now starts at Tatoosh Island. It used to come clear into Neah Bay, and past Neah Bay clear down to the Sekiu River. That was considered ocean fisheries. They propose now that we get the [ocean] fish from Tatoosh out. This fellow I had trouble with, I think he was registered Olympia. It was a very small boat, and it probably hurt him very much to be cut off—really. He was a young man and hotheaded. He'll learn to take things in his stride when he gets older.

But you see the sportfishing last year was delayed. The commercial boats started two weeks in the ocean before the sportboats could. But none of the . . . sportfishermen and the charter boats complained. I think Ilwaco and Westport have more trouble with the commercial and the charter boats than we have here. [It's because there's] more of them, three or four hundred charter boats in each place. We only have . . . eight or nine charter boats at Neah Bay and only two or three here [in Port Angeles]. We fish differently, too. They drift down there in the ocean; I can't imagine how many people must get sick drifting. But we keep our boats moving. We only get maybe two or three in a whole week that might get sick.

In the spring when I go up there [to Neah Bay] for my charters [before the ocean is open], I have to stay inside of Koitlah Point. But I'm not going up in April this year. I'm going to go May first. By then I should be able to fish ocean fisheries. There's fish in the straits there, but the Indians now have quite a fleet, and they fish in there. Sportfishing is taking fish away from them. If I wait until I can get out into the ocean, then I won't be taking their fish. I wasn't requested [to wait]. I can't get out [if it's rough]; then I have to fish in there with the Indians. The Indians fish in there because it's close to home; they're five to six minutes from the dock. [Their boats are] thirty-six, thirty-eight feet—some of them are bigger. They troll.

Retrospectives and Prospects for the Future

Mars and Gaylord have three daughters and one son. Their youngest daughter was born in 1952. They did not take their children kelping with them, but in later years, when Mars took her charter boats out during the summers, some of her children went with her.

I had an older daughter—the number one. Then there was a little space— six years—between them. She was a good babysitter. But we didn't leave them too long unless we had an aunt with them or his [Gaylord's] mother. When the older girl was working—why, the next to oldest was with us, baby- sitting with the other two. When she got old enough to work, then the boy went with [my husband], and the little one went with me. She was kind of a Mama's girl. She was with me all the time; she was my automatic pilot. I would refer [my son] to his dad, because the male child has to be told by the male in the family what's expected of him. The mother has to kind of feel him out to see how arrogant he is. Only one relative [ever criticized us for taking the children]. We ignored it.

I think [fishing] is a clean way of raising [children]. We [had moved] away from the city to Bainbridge Island because we wanted our kids to get a little more rural life. [In fishing, the children] are working too, with you, and they're earning money from the time they're little. They learn to earn it, and they can learn to spend it. They all came out good that way. They know how to manage their affairs now. It was good for them, I think.

Mars Jones is part of a very close community of family and friends, both on and off the water. Her son and one son-in-law are also involved in fishing and chartering. Her son fishes with his father in the summertime, when he is not teaching. Her son-in-law recently set up a skin-diving charter business with Mars's and Gaylord's assistance. Mars speaks with concern about the past and the future for their way of life.

I've got all kinds of friends—in the music business, and my children's friends, and . . . commercial fishermen and charter boat operators. Then I've got all these people [that I take fishing]. Some of them come from all over the world.

When I get time I paint—watercolor and oil. I've always made bread, and I love to cook. [I don't play music] any more. I garden. I have a garden every year and go off and leave it to go fishing. We get a lot of produce out of it. I get home just enough the early part of the season [to] keep the weeds down. My big problem is to get somebody to water it. The gardener comes and waters our lawn, and I keep telling him every year, "Please water the garden. Forget the lawn."

[My husband's] probably the reason I'm doing this [fishing], because he encouraged me. [We have an equalized kind of marriage], but we don't have women's lib. He comes home from a trip and puts his feet up and waits for dinner—that's normal. Because I feel like . . . in some ways we still are women, and our place is . . . a little below the man, no matter how you look at it. That's the way we were made; God made Adam first, and then he made Eve, and if you don't accept it young in life, you will later. That is, if you love your guy.

I never got lost [at sea]. One of the reasons that I felt like I could do it from the beginning was that he used to take a nap—or he had a headache—and it was up to me to find my way in. And that was good for me, because . . . when we were first married, he wouldn't let me drive the car. He was afraid I might have an accident . . . and our insurance would go up. In fact before I ever . . . got a driver's license, I had my boat-operator's license. Can you imagine that?

[My son] said when he was little that he was going to be a schoolteacher so he could fish in the summer—and that's what he did. [He fishes with his dad.] [We are a pretty close-knit family.] We involved the kids in [fishing] and they profited by it. We never wanted to be rich; we just wanted to be together. We have family get-togethers. And . . . every time there's a new boat in the family, they're all right there; it's like having a new baby. Our son-in-law, number two daughter['s husband] . . . he's a charter boat operator and skin diver. They are going to base here in Port Angeles because he wants to be near the kids. He's got our third boat *Tracy Jo,* and Tracy Jo is our newest granddaughter. We just got them that boat last November.

[We have] three boats: *Satin Doll I, Satin Doll II,* and *Tracy Jo.* In fact, we had the first fiberglass boat . . . up here—*Satin Doll I.* The fishermen just started getting interested in that, and gradually [the company] Blue Fin built this little Hoquiam Boat Shop and builds these hulls. There must be five or six of them in this harbor now, but we were the first fiberglass boat up here. It saves on your maintenance; it's really fantastic.

[I expect to continue fishing] probably another ten years, I hope. Eight years, anyway. I hate to say this, but I think that sportfishing will probably last longer than commercial fishing. They're going to be getting into fish farms; I think that's going to be the thing of the future. In fact, if you want to buy a fish or eat a fish, you're going to have to buy it from a fish farm. [I think the quality of it] will probably be very poor; your hatchery fish are more prone to disease than your native fish. I think [there'll still be fish out there to catch] if we're careful. The most important thing would be to get everybody out of the rivers and let those fish do their own thing. But how can you get anybody out of the river? I heard Oregon lets commercial boats—the little kelpers—fish right in some of their rivers. Yet they're the ones that want to close the Columbia River. I just don't understand it. If they took all commercial fishing off the mouth of the Columbia River—say, maybe three miles or something. . . . If they're hard up for fish in the Columbia—they might get more fish back. And they'd let nobody, not even the Indians, catch them in the river until we build them up. But I know the Indians . . . won't go for that. If they do it just even two years, and nobody catches in the river, it might help. [But boat payments are a] terrible [problem. I know someone who bought] a brand new boat this year. How is that young man going to pay for it? The season is so short, and most of them had to sell their old boats, and some of them haven't even sold their old boats yet. Who wants to buy them?

The gear buy-back program was all right until they turned around and sold [the gear] to the Indians; that really hurt the white man. I [feel] about the Boldt Decision just like anybody else. It was a wrong decision. It made a lot of animosities between the white man and the Indian that need not have been. Before that, the white man and the Indian were getting along good.

We have a lot of Indian friends, and they're *good* friends. They're hard-working people when they're doing something they like. And I like to see them out there fishing, because they love it just like these white men here. If he's a fisherman, he's a fisherman! And it doesn't make sense to stir up so much animosity. It isn't the Indian and the white man against each other, because they used to get along. [It's more a problem with government regulations and big business].

The industry's been mismanaged. Our Washington [Department of] Fisheries was doing so good. We were the first place in the world that was doing the fish hatchery thing and putting fish back into the water. They saw ahead—trouble ahead—many years ago. This fellow that was head of fisheries a number of years back—he had some saltwater rearing ponds—he was way ahead. Then he went out of office. They let those fish—the fish came back to spawn, and there was no place for them in those saltwater ponds. That was terrible. And I don't believe the head of the fisheries department should be appointed by the governor. I think it should be somebody that the fishermen themselves elect, somebody who knows about these different branches of the fishing department. Because just putting in any old body . . . they're at a loss; they don't know how to manage . . . the fish-management program at all.

Marya Moses

Marya Moses

*Marya Moses, a Tulalip Indian, was born in 1911. She was one of
fifteen children and fondly remembers her early years and traditional
Tulalip fishing practices.*

Family and Community Fishing Traditions

Mom was eighty-three or eighty-four [when she died]. She had fifteen
children [three daughters, twelve sons] and about five miscarriages. [Growing
up in a big family] was nice. We got along real well. We all loved each other.
I don't remember fighting even. I suppose it was our mother. I thought from
the time I was little until up till I was thirteen [that] she was an angel. The
old people would come . . . [walking] and spend two, three days with us.
And Mom would get out her berries, and then they would tell us Indian stories [around] that long old table. I liked old people. Their values were different. They looked at you for the way you were. They could see—they must
have seen what was in you. There's no old people, just me, now. I remember
lots of stuff [from] way back—*ahn-ka-tee*—that's the Chinook for "long ago."

[I can] see that old kettle yet. [Mom] used to cook beans outside on the
iron kettle that had three little legs. She'd start ladling it out for us. We
didn't serve ourselves; she served us. That way everybody got an equal share.
She couldn't read or write, but even townspeople loved her. She was just
different than I was. She was just kind and considerate. You would even put
your arm around [her]. Not like me; I said I was always different. [Around
age thirteen] I started going boy crazy, and she started stopping me, and I got
so I didn't like her. Then I think we all kind of started going different ways.
But we got along real good, as crowded as we were.

[Mother] had a hard life. [She] had to make sure we were fed and housed
and everything, because Pa didn't. We used to sleep four to a bed—feather
bed, mind you. We were so poor that we just had a board bed. We lived on
ducks, and [Mother] saved the feathers [for our beds]. We all had feather
beds and pillows on the floor. She used to get a little bit of stumpage money.
There's this mistaken idea [that] Indians get money from the government,
which they don't. They were supposed to, but that isn't true. The Bureau of
Indian Affairs [let them log] her property, and that forester took 7 percent
of that money. So they weren't doing Indians a favor. They were doing
themselves [a favor]. They were getting paid plus taking a percentage of the
money. She used to get about $50 or $75 a month, which was quite a bit at
that time. Everything was so cheap. But I would say *she* made the home, *she*
held us together. Even after she was old, she *still* held us together. I don't
know if that was good or bad. The only one that broke away from us is my
oldest sister. She's in Montana. [The rest of us stayed here.]

Mother [became] senile the last three years. That was the hardest. All the
kids and them others, they wanted to put her in a home, [but I said no]. I
had to wash her, clean her up, and so. The preacher here came with just
prayers—which didn't do anything for her empty stomach. He's dead now—
said I wasn't taking good care of her. And the social workers came and they
said, "She's on welfare." I said, "I'm not getting welfare for her. You go ahead.

Take your few paltry dollars. She can eat what I do." When the nurse came, I said, "What have you come with? Any medicine?" See, I am ornery. You can judge for yourself. I said, "You took a vow to relieve the suffering of humanity. Now, what have you done to justify that?" [When the nurse did not bring medicine], I said to her, "What the hell are you wanting here, then? You're taking up my time. And if you ain't got something to contribute, don't bother me." Wasn't that awful, the way I talked? But I was under pressure. I was working and taking care of Mom, with kids in school. You wouldn't want it. Nobody seen that part of me, though. They just said I was mean.

My [father] lived to be eighty-three. I think if he didn't drink so much, he would have lived longer. He went on a drunk for three weeks, and then when he come out of it, he had heart failure.

[In the old days] all Indians fished [women and men]. They used to make their own nets [and canoes]. They caught [fish] by the tons. If [a woman's] husband got sick, she'd have set the net, a little net. They'd just have a canoe. They would go out so far and stake it. There was no anchor at that time. Grandpa Moses would make a homemade trap. They made a trap where the fish would go in and couldn't get out. The women did all the work. They caught the fish and cleaned it, they dried it, and got their own wood half the time [and then took care of the kids]. Well, that's the old [way. And] them old chiefs sat around and talked about their past glory.

We never wasted anything in that time. They ate nearly all parts of the fish, most parts of it. They told you if you ate the tail of the fish you would marry a chief. They always had little stories. They had hard times long ago. They didn't even waste the spine. You know, when they got through eating the salmon, they'd put the backbone in the basket. Later on, maybe in early spring when times really got tough—you know, freezing—they would make a soup out of that.

[The Indians] lived on fish and ducks. Ohh, the ducks! There were so many that they used to have little traps for them. They'd weave stuff and put it over the marsh. [They'd] clap, and the ducks would fly up and into it. Then the clams—oh, they were plentiful! They watched the moon. [Then] they'd go over to them islands over there [Camano, Whidbey, and Gedney]. The women would fix a lunch; then they would go and dry the clams. They'd spend two or three weeks over there drying the clams and preparing for winter. They strung the clams with the cedar inner bark. [There] was just so much around. They dried clams and fish, and they had berries, deers, ducks. The reasons for putting away the clams were for winter use and for trade. They traded dried clams and fish to each other and to other tribes. For example, the Yakima would come down and get dried fish for some blankets, beads, and other things. When the white man came they used to go and trade there in Everett. They would holler "Clams. Clams. Two bits."—that is what my mother told me. They traded to each other, too. They always sold [fish], as far as I know. I mean barter—that's something like selling.

I would say that [for] Indians, since the beginning of time, [fishing] has been important. It was their whole living. The fishing was so important that when the white man made a treaty with us, we asked nothing but to keep our fish, clams, and berries. We let all the other, like timber, gold, silver, coal, and everything [go]. So you could see how important this fishing was to the Indians here. We asked very, very little in return. Now, it makes me boiling mad when they made it sound as if it is greed when we insist on our fishing rights. The greed is not on our part.

Early Years in Fishing

Marya attended the Tulalip Boarding School, where she says she was always "asking questions and being mouthy." She married and had eleven children. When her oldest child was twenty-one and her youngest was one, her husband died. She got into fishing through one of the many odd jobs she held to "supplement my meager income from welfare."

[I have] done a little bit of everything. I would say that Indian women, they all . . . worked at home. Like me, I knit and trade or sell the Indian socks. When I wasn't fishing, I was cutting shake boards. I learned [how to cut shake boards] from working . . . alongside my husband. It was a blessing in disguise, because when he was gone—why, everything turned out wonderful. I got off welfare.

I didn't have a home. Really, I didn't have anything. I had so many children no one wanted to rent to me. After he died, things changed slowly. But I had to work at it. It didn't change by itself. I had to make it change for the good—make a home. I worked in a cannery when my children were little. I didn't work there long—about six months, I believe . . . because my girls were at that age where the boys ———. I think I'd rather go a little hungry than my daughters go ———. So, I quit.

[I got into fishing] in '52, I believe. I was left with nine children and two in the service. I'd had two operations when my . . . two sons were in Korea. The first one was in September, and the second one was in October. I couldn't work then. I was on welfare, and seemed like they were always breathing down my back; every other day the gal from ADC [*government Aid to Dependent Children—ED.*] came for a visit. I don't like anyone to do that to me; I'm kind of an independent person. I don't like to be under obligations to anybody. So when I got stronger, I had a few visitors. They always came, and they'd eat and stay for two, three days.

I got a job cooking for my nephew. [It was] the only thing I could do, because I have no education and all these nine children. Oh, I could start cooking on the beach for them fishermen, and they said they'd give me $15 each a week. [I was cooking for] a crew, four or five. Oh, they could eat! Oh, criminy! For breakfast, some of them would eat two eggs, some four eggs. One of those guys ate six for breakfast. The crew drank coffee all day long. My children and me, we'd get up four o'clock in the morning and make coffee. Then I had the girls, four of them. They would help do the little dishes and [would] pack water from quite a ways to rinse those cups out.

In about two weeks they didn't catch any fish, and I was supposed to get paid each week. My bill ran up so high that this woman, Mrs. Stiles, told me, she said, "Marya, your bill is $300." I said, "Well, I better look for a job." Mrs. Stiles owned a store on the reservation. She was a kind woman. My bill ran up so high that I said, "I can't cook anymore." Four of them paid me once, and what is $60 for two weeks? So I said, "We're going home." And Tom, my nephew, said, "Hey, Marya, why don't you fish?" He took a little piece of net, about four hundred [feet] and said, "You can have that." So Neil hung it for me and we went home. I was really worried about the bill.

Then, a junkman came along, and he said, "Can I buy your old batteries or anything?" I said, "Sure, help yourself." So he looked around, and I told him my little story, and he was real good. He said, "Say, I got an old boat that got wrecked . . . in a storm. It ended up over in Camano, and I had to chop off two feet." He said, "It's good, and I'll let you have it for $150." I still don't know where I got the $150. But he took some of my junk and whatever he wanted, and then we kind of horse-traded. Well, I got that little sixteen-foot boat. It had a little teeny engine [and I had] that little net. So the girls and I [started fishing]; my crew was my daughters. Vicki—she's a nurse now at General [Hospital]—she was the splash girl [and] . . . she was good. Rachel and April . . . had the winches. I had a woman crew. They used to say, "Oh, Marya and her all-girl crew." Also, Delbert Moses was bull wincher.

The men used to go way out and drop their net—quite a ways out; they'd drop their net and come in slow. I thought, "Well, gee. They're way out there. Wouldn't take me long to drop in right close." And I'd come around quick. And here I was corking them in a set. They weren't catching any fish, so they said, "Make a southie." Southie's opposite what the men were doing. So I said, "Southie? Huh uh." And I thought, "Southie? Well, you go south, I guess." But really you were supposed to go the other way. So I made the wrong set, and that's the way the fish were coming—and oh, man! Did I catch the fish! But from there it just kept right on. [When I made the wrong set] they come running out [and] said, . . . "Which way'd you make a set?" I thought, "Hell with you guys. I won't tell you!"

After that they were often kidding me. They'd say, "Let her go, Marya!" and here it'd be a riptide. I didn't know that then. But sometimes it backfired on them. One time at that place over Mission Bar—where them pilings are sticking up—they hollered, "Let her go, Marya!" The tide was running real fast. I really believed them. I went out there, and the net wrapped around the pilings, and my nephew Cecil started cussing. Getting more stubborn, I went out again, and they were laughing. But we caught a whole lot. Boy, you ought to see them move then!

I learned to go way up, enough to compensate for [the tide]. I would be able to make a short set. The men's net was eight hundred feet long. I was able to make three or four complete sets. So, I often corked them three or four times. I didn't know that I was doing that—after all, it's not fair. I made banana sets. I would go right out and come right back in, but I would still catch them. Boy, I started nailing them.

Of course, we worked there. We didn't get rich, but I was able to pay a few of my bills. And then I was able to get off welfare, too. We didn't make that much—like 500 pounds, where them guys would make tons. But that was a lot to me if I made 700 pounds. We only got two bits a pound; we didn't get very much. But them other guys would make tons—like Tommy would make four tons. But I was so thankful for even that little bit of [fish].

I was very, very thankful. I also didn't think it was my ability or anything out of me. I always thank God for everything. One person made fun of me. "Do you ever watch her? She talks to herself." But I was praying—I'd be saying "Our Father." I never told this, and I'll tell it now. I would say a prayer. I didn't know I'd move my lips. You wouldn't believe it the way I'd also get up there and curse some of them. No kidding. I didn't want them to know what I was doing. It don't matter now. I don't know how [I did it sometimes]; it must've been God that helped me, because my net would be up, but still I'd catch the kings.

Then a man [who had] had a stroke sent for me. [He] said, "I'll sell you my boat and net and everything for $575." He built the boat himself. So I got that [for beach seining], and that's what I'm still fishing with today. It was originally twenty-six feet, but we had to chop it down to about twenty-two. But at that time the price of fish was only eleven cents for chums, and you know, it's quite high now.

Recent Years

Marya expresses love for fishing. She and her family own several boats: her beach seine boat named St. Anne, *a set-net boat, and a gillnet boat* Four Winds. *Two of her sons operate the gillnetter, and her daughters set net.*

I just love fishing. Why, I could be so depressed and be laying there sick with a fever, and I'd start thinking about summer and see the beach and the sea gulls. I could just see it in my imagination. I could see a jumper. Sometime I'd be sitting on a beach having a cup of coffee, and they'd see a jumper. And as old as I am, you'd see me try to run and scramble onto the boat and get out there. To some people [fishing] might be hard, but to me it isn't hard work. I love it. Even if you offered me a job [that] paid me thousands, I wouldn't take it. I'm not kidding you. I just love it—it's too bad I couldn't be buried at sea.

I like the beach seining. That's my life. I take no part in the gillnet. [It's my boat, but] my sons Gilbert and Danny run it. It's just a twenty-six-foot boat [and they fish] the San Juan Islands, up around there. [The set nets] aren't allowed outside, only in the inner bay [of Tulalip Bay]. I like set netting, but my bones are—[well], I have to take care of myself. I get cold; my ankle'll hurt, or something like that. It's a lot of fun, and it's a lot of work, too. It's mostly checking your net. You have to check your net, and your hands are in the water all the time. It's too cold to be out there. Beach seining is my life. I love it. It's not so much work to get out there.

[Now, I have] four [helping me with seining]. There's a splash boy, three men on the winches, and myself. I didn't want to pull too much this year. I wanted to mostly relax a little bit, just be out there. I had pulled when it was necessary. Years before, I used to pull right along with the men. I would take the corkline, but I didn't pull on the leadline. This year, I didn't do any of that. My grandchildren are getting old enough. I've got a lot of grandchildren, but I pick out those that really like fishing . . . [and] relieve those that aren't really interested in fishing.

The crew that I pick, they're handpicked. They have to know what they're doing. They've got to take orders. If I say, "Roll it in—fast," or [use] signals [they do it]. They do as they're told. There's four of us. Each one watches for the fish. I've a grandson. He's into it now. Then another little grandson. He will be a splash by next year.

[Here's what I do.] Here's a boat. See, this is a boat now. I go out [from the shore] like that. There's a net now. I'll make a kind of C. I make that bend like that [and the splash], he stays right there and splashes. The fish will come in this way. [Without the splash], they would funnel out like that. But he's right here, and he splashes, and [the fish] come here and they'll go back. He has these paddles—or oars, rather—and he has to keep that up all the time. You hold about five minutes. Ten minutes is the longest you hold your set; that's old tribal regulation. It gives the next guy a chance.

[Now for] beach seining, you have to buy a permit from the tribe. Last year it was $35 or $50. And you get a location—it's six hundred feet—and that's where you fish. I think we are allowed, they said, from mean tide. You know, we own the beach front—all the so many feet there and then so far out. That's where we fish. [The fish will] pass. See, you watch out there for jumpers, or else you could make a blind set; you just go out there and try it. But the fishing hasn't been that good; I don't know what's the matter.

The tribes make their own rules. Each tribe has their own, and the people have a say to that. [At the last general council meeting] the people voted that just a tribal member can fish. But this year, I hope to have it changed so the spouse can fish. They have to be married and live on the reservation, if I have anything to say about it. Now, take my daughter that's married to an Oklahoma. She just had a baby, and within a week she had to go and tend to her net, or else have it picked up and she would lose it. Her husband could run the boat, but he could not touch the net . . . and the patrol men were watching too. This is the first year that I know of them to get that strict. I think they should be. I don't think you should make laws and break them, just because it's your relative. I've always said that. That goes for the white man too. And, what's the use of having laws if you don't enforce [them]?

I wasn't able to make any money this year [1980], because the damn old state kept shutting us down. Of course, you probably hear them cussing us, too. [I fished] maybe two, three [days this year]. There were game wardens standing right over us, making us turn back the fish—whatever species they wanted. That was hard. If we killed any or caused them to die, they would shut us down. They were right on the fish buyer's boat. If you caught too many [or didn't] follow what [you] were supposed to [i.e., the regulations], they shut [you] down. Why, I'd start out [like] today, and pretty soon they would come out: "It's closed, Marya"—or something like that.

And now I feel as if we are being denied this one [and] only thing that we ever asked, and we are made to sound as if we're culprits—greedy, diminishing it, and all that. Where I feel as if it's just certain people like them sportsmen, you know. I don't know if you've ever seen it. I have. I've hauled in fish with the mouth ripped open, where them fellows get out there, and they'll hook it, [then] they throw it back. Them poor things. Sometimes it's in the eye. No. What gets me is how we've never asked anything. And what is the fuss? Long ago they used to sneer at us "fish-eaters!" It was almost a disgrace, because in them times . . . they looked down at you. Now it's real expensive.

Marya is concerned about the effects on fishing of increased land development and the thoughtlessness of the new residents on and near the Tulalip Reservation who have moved here from Seattle.

[When I first started fishing, things were different.] The [white] people that lived here all the time, they were real good. They knew what they were getting into. They were living on the reservation and they didn't mind it. They liked the wild, the trees, and most liked the same things we do. They bought land here years ago for ten, fifteen, twenty dollars an acre. They've made their homes here and they like it here. The third generation—like my daughter's generation—they started selling land and they sold it cheap. That's what you see building up now. Now, a lot of the [rich ones from Seattle] have homes here. They come here, and they have skis and power boats, and ohhh . . . that makes a lot of difference. They could scare the fish right away from you. They know it, too. Some of them [have] even run over my net and cut it. There's no sense to it. Here's a beach like this, and we are allowed out so much. And here's a whole bay, and they come around like that [on purpose], at full speed. You know, them Boeing workers and different ones [with] high-paying jobs have great big boats, and their little brats have them speed boats. And oh, that bay looks like a town, there's so many [boats]. Of course they could afford it; they're making a lot of money. I hate to see any more of them come here.

Commitment to Fishing

There is [money to be made], but you really have to work at it. You have to live and breathe it. You make money, [but it] goes right back into it. Now, you'll notice there's a light on that gillnet boat. We give it a light all night— night and day . . . [to] make sure she don't get damp inside. Four thousand dollars went right back into that [gillnet boat] for next year. Now I'm down, my money's down again. You can't just go out there and make money. You have to watch your equipment. Last year, I bought nets—over $6,000. That isn't the hanging twine; [so] you pay a man for hanging it—$500, some $400. You make money, but you put out a lot, and you're thinking about it all the time.

I find my children are coming to it, even the ones that didn't like it. There's one daughter—she wouldn't even eat fish, but now I see she's got her own little boat. She's working up to it. And when they say, "Mom, can I have corks? Can I have this or that?," I say, "Listen, I worked and earned everything I got. If I help you ———." You know, it isn't good to give kids everything. Let them earn it like I did. Now that sounds kind of mean, but . . . it's a cold world out there. You have to be able to stand up and can't be fainting everytime somebody speaks rough to you and all that. Maybe now [the children] are bragging because they had to work. I think they sound proud.

Two of my [daughters] have gone into it [and two sons]. They even learned to hang the net; they have their own net, and that's something. They got a little teeny boat and they set net. See, they're just starting out like I did. I don't want to deny them of all the enjoyment. [Finding out for themselves], that's the best part of it. They'll be so proud when they get old and look back on it.

[My daughters] married Oklahoma Plains Indians, and you'd be surprised, them men are just so interested in it. I tell them, "Don't go into fishing. If you can find another job, do that." I wouldn't advise anyone to, if you're going in for the money. My aunt Celia fished. She owns a boat; my grandniece goes out there—they have a little boat. There's some that set net, quite a few. Lots of other women fished. But there was a lot of them that just [tried] to get in it for the money . . . that's what I think. They didn't like fishing like I did. I really liked it.

[Some of our women are] married to whites. [The women bought boats so they] and their children [could fish in our areas]. But they didn't take care of their equipment, I'm sorry to say. Their kids didn't take care of the equip-ment [either] and the boats just went. They weren't really true fishermen. When you are, you take care of your boat; you take care of your net. That's where your money is coming in. A person—if you're into something, you should see to it. I see to it that the man that skippers my gillnet knows about it. I wouldn't just put anyone [in charge] because I like them. I take care of [the boat] because I like it. [Your boat is part of you.] *St. Anne* is part of me. We just love her. I was going to sink her this year, and Neil said, "No. I'll plant flowers around her." Now *Four Winds*, she's become part of us [too].

Tulalip fishermen are swell guys. You find this out when you break down. They loan you parts—propellers, spark plugs—and give free advice; what-ever they can do. Up in the straits, around the San Juan Islands, the gill-netters watch out for each other, especially if a storm is brewing. When we're beach seining at the [Mission] Bar, friendships cease; we're competitive. When we're gillnetting, it's different. We look out for each other; we're friends.

Helen Gau

Helen Gau

Helen Gau was born in the mid 1920s and was raised on a farm in eastern Washington. She was one of eight children.

Childhood in Eastern Washington

[My mother] came here from England when she was nineteen years old.
I'm sure she left her true love in England. [She] cried all the way across on
the boat . . . my aunt tells me. My mother was a ring-tailed monkey, but
my grandmother was something else beyond that. Both these women were
strong and aggressive because they had to be—to survive.

I'm sure that my mom really married my dad to . . . get away from my
grandma. My dad was a very Bible-reading, very devout man. My mom was
a fun-loving person. She loved to travel. But my dad was strictly a stay-at-
home. It's just too bad there hadn't been such a thing as birth control in
those days, because to have eight children and to be so frustrated as my mother
must have been. Not that my dad wasn't a good man. He was. But he was
willing to settle for mediocrity. I'm sure if Mom had had her way, she'd have
done a lot more things in her life. [It] took me years and years to realize that.
I thought my dad was the perfect one. She really wanted things for us and
wanted things for herself, and there just never was enough money on the
farm. Dad couldn't have cared less; he had his horses and his fields. He got
up every Sunday morning and walked through the fields, and that was his
joy. But the fun things and the interest in life—if any of [the] eight of us
have it—came from my mother.

I look back through my life, and the only thing I ever remember [is] my
dad and mother quarreled constantly. It was just total uproar in our family.
Never any physical violence, but just this mouth all the time. My mother
was always berating him. The only thing that I ever remember these two
people agreeing on—and I have a tear in my eye and in my throat as I say
this—was telling me, at age fifteen, "Either shut your mouth or get out." My
dad said this to me, my dad that I dearly loved. Six months before that, I'd
gone to him and said, "Hey, look. There isn't enough money." My dad had
been laid up with a bone problem for a year. I was starting high school. I
wanted some clothes, and there wasn't any money. I'd asked them, "Couldn't
I move away from home and find a job working?" They wouldn't permit that.
So then this big, hairy family uproar happened, and they said, "Either [you]
be quiet and live the way we want you to live, or you leave—or I'll call the
sheriff." In those days you went to the sheriff. He started for the phone. I
said, "You don't have to do that. I'll go."

I really know—and this only came to me about a year and a half ago—
that [telling me to be quiet or leave] was the best thing those people could
have ever done for me. I really think my mother knew exactly what she was
doing, in that she did not want me to have the kind of life she had. She
knew I was strong-willed enough that I probably would go off and just do all
right. Not that I wouldn't have problems, because I would. Dad probably
didn't perceive it in that manner. But it's such a thing in my mind that Mom

knew exactly what she was doing. I only really have one regret, and that is that I can't sit and talk to her now and say, "Hey, I really understand." But when she was dying, I was the one that was with her. She knew I was there, and she knew I was caring.

So I left home, and I worked [at] just about anything that you wanted. Anything to keep body and soul together. I worked my way through high school by room-and-board type of thing. Then I'd work on the apples . . . in eastern Washington. I'd thin apples. I'd pack apples. I'd do whatever I could. I found out in the first week the thing that I wish every young child, at fifteen, could find out. There [is] nobody—but nobody—that'll ever love you like your mom and dad. Bar none. I also learned another thing in that first week. You either keep your mouth shut, or you're not going to have a place to eat or sleep one of these times. To me, it was really the making of me. Basically [it was] the kind of thing that really sets up a character that you're going to have for the rest of your life: the determination to make your own way.

Then it came time to graduate, and there wasn't any money for pictures, and all this sort of thing. So, I didn't graduate from high school. I was too proud to say, "I don't have the money for this." So, I just walked off from high school. It's a hurt that hurt me a long, long time. But [later] my two girls saw me graduate from high school. [I] guess that's why I don't like to see people putting young people down, because you really don't know what's going on. We do some terrible things to kids without even realizing.

The Search for a Boat

When she was twenty-four, Helen married Keane Gau. She became mother to his three-year-old daughter from his first marriage. Then they had another daughter and, later, a son. Helen and her husband operated a tavern in Spokane, but they had outdoor interests. They were avid sportfishermen and they read about long-distance sailing. At the end of 1955, Helen and her husband decided to give up the tavern and try the unusual combination of sailing and trolling.

This is really the way it happened. We came to the coast on a vacation. We'd never been on salt water, either one of us, in a boat. But we came over here and threw our boat in the [Puget] Sound down here and took off with our two little girls. [We] went down to La Push, out in the ocean. I shudder when I think about it, because we passed Tatoosh Island with the storm flags flying . . . which means there's a storm imminent. We had no idea what it meant. That's how green we were—and how stupid, really. There's a saying among the fleet that God looks after fools and fishermen. It's really true, because we went down that coast with those two little girls . . . both flags were flying . . . and had no problem.

We spent a month on that vacation. That was in September when we went back to Spokane. When I got home I realized I was pregnant, and we wished there was something we could do. I really wasn't happy with the tavern business. My husband's always been an outdoor person. So we just started kicking it around. I laugh now, because on Christmas Day we hadn't even really made up our minds to do it. On New Year's Eve we told our friends and family, "Hey, we're selling. We're selling out. We're going to buy a sailboat, and we're going to fish." That quickly we decided we were going to do it.

Everybody thought we'd absolutely lost our minds because we'd left so-called security. But his grandmother, who was well up in her seventies then, handed us this book by Ballard Hadman.[1] She wrote one of the best books on trolling that I've ever seen. Grandma handed my husband that book. She said, "Take this, Keane, and do it. You guys'll do fine." She was the only one that really gave us any encouragement in his family. My family said, "Well, do it," because my family were farmers and had no fear of stepping out. But his dad had always been in business and thought that's where we should remain.

[We] sold everything we had and bought a mobile home and bought the boat. [We] went looking for it from Spokane to Portland, Oregon, and every puddle of water in between. [We] finally found a boat in Tacoma. We kind of had an idea of what we wanted. We wanted enough room for three kids. We wanted a boat that really would sail around the world some day. We happened to be visiting some friends in Tacoma, and we hadn't found a thing. We knew we were doing something that hadn't been done by getting a sailboat. We were absolutely sure of that. Because nothing we read or could find out ever indicated any salmon fishing was being done off a sailboat. But that's what we were after. We visited some friends, and their oil man happened to come in. This friend . . . said, "These people are looking for a sailboat." He kind of described what we were after, and [the oil man] said, "I know where there's one down near Fife."

We had probably looked at close to a hundred sailboats by that time, and each time it just wasn't what we wanted. I stayed in the car, because I was about seven-months pregnant by this time. [Keane] went walking down to look at this last boat. We had been kind of down in the dumps, because we hadn't found exactly what we wanted. He came walking back and he said, "How does the name 'Eileen O'Farrell' strike you?" I said, "Fine." He said, "C'mon. I think we found it." And we knew it. And incidentally, when it came to buying our second sailboat, the same thing happened. The minute we stepped on board, we knew. We just looked at each other, nodded our heads, said, "Yeah, this is the one." Just as simple as that.

[We chose trolling] because—well, hell, when we first started out, we didn't even know one thing from another. But it just looked like the kind of life

1. Ballard Hadman, *As the Sailor Loves the Sea* (New York: Harper, 1951).

that we'd be interested in. It wasn't any night fishing. Gillnetters do fish at night, and that didn't really appeal to my husband and me. Net fishing didn't either. We were avid sportfishermen. That's probably the real reason. They're close to trollers. We look back now and say we had unmitigated gall.

Family Fishing

For their first season, the Gaus fished off of Neah Bay. Most of what they knew about commercial fishing came from books; so they had a lot to learn.

The boy was born on the third of May. He was five weeks old when we left . . . Spokane with our mobile home and pulled into Neah Bay. We day-fished out of Neah Bay that first season. I didn't go every day. I did go enough to know that I really liked this. [But] I was scared to death that the ocean was something wild, as far as I was concerned, with a baby on board.

The first time in Neah Bay, I hadn't gone out with my husband that day. Everything was set up. I was to watch, and when the boat came in that evening, I was to . . . walk out. If he blew the horn, that meant he had fish, and I was to go down and see his load. So he came in and blew the horn. I took off, running with the baby in my arms, down to the fish house. I got about halfway down there, and I suddenly realized the boat had stopped out in the middle of the bay there—Neah Bay. [It] wasn't coming. I went over to the fish house, and I thought, "What's going on?" I couldn't figure it out. Pretty soon he came into the dock, and I looked down. I could see that he had fish on board. But he had a face. [He] really looked down. I said, "See you got fish!" I was just elated. There wasn't really enough fish to buy a sack full of groceries. But I was elated he'd caught some fish, and I thought he would be, too. But he just really looked beaten. He said, "Yeah, I caught some fish, but I sure blew it." Our poles, on the fish boat, stick out from the side fifty feet. He'd forgotten to pull his poles when he came in. He came steering right through the fleet and hit another boat with our poles. He tore that fellow's poles off. He tore our poles off. He broke the window on the pilot house. The old fellow was in bed, and he came up on deck in his underwear—just wild. Keane apologized and said, "Gosh. I really blew it. I'm sorry." But the guy was really reasonable. He said, "Well, it's okay." We said, "We'll go get you another pole and fix your boat and everything." He was really, really reasonable about it, because he could . . . [have] charged us for the time he lost fishing, and everything else. But he was a great guy, and he didn't. We aren't the only boat that's done this. It happens quite frequently.

The following week—our little daughter was six years old at the time—and she'd go out. She'd stand on the beach and she'd watch her daddy come

in. He used to wait—most of the fishermen did—till they got inside the island, so that it wouldn't be so rough. Then they'd pull their poles. But she'd
run out there, and she'd wave her arms up and down, "Daddy, Daddy! Your
poles! Your poles!" Trying to tell her dad to pull the poles. I don't think
there's a mistake ever been made in the salmon industry we haven't made.
But those are the kinds of lessons that are really valuable.

Then we happened to meet an Alaska fisherman and his wife who were
really great people. They said, "If you're going to fish with your family, why
don't you consider Alaska? There's quite a few people up there doing it. It's
just a whole lot easier on families because there are harbors. If you don't want
to trip-fish, you can sell your fish every day to a buying boat." This wasn't
possible on the Washington coast. So we said, "Well, we'd like to think about
. . . [it], because this is why we bought this kind of boat. This is what we
[had] really wanted to do, was to have our family with us and go." So, we set
our sights on that the following spring.

We worked like dogs all winter and got our boat [in shape]. We'd put a
wheelhouse on it so that we were in[side]. The first year we were right out in
the weather. We got down to the wire; we made our decision. We were going
to Alaska, and the last thing we had to do was haul the boat out and bottom-
paint it. By this time, of course. . . . money was really getting scarce. [My
husband] said, "You come down and give me a hand with the bottom-painting." Well, I drove in, all set to give him a hand with the painting [because]
within a few days we were leaving for Alaska. And there he stood, just looking sick, sick, sick. I looked at the boat, and it looked like half of the bottom
was gone—dry rot. It's when the wood rots, and the only thing you can do
is to take [it] out [and] replace it. We stood there, and he was just non-
plussed. He said, "Well, shall we be sensible and forget about going to Alaska?
Or shall we bow our necks and spend our last 500 bucks and fix this and go
anyway?" I said, "Well, we've come this far. Let's go." We had about 600
bucks left, and the bill came to . . . right around $500. But we already had
our groceries, and everything was ready. So we just fixed the boat . . . [and]
we went.

We got to Ketchikan. I wonder how, because on the way up [my husband]
was overcome with carbon monoxide. I thought he was dead for about an
hour and a half in the Queen Charlottes. [It was] just a real fluke of a thing.
He had redone the exhaust system. He brought the exhaust system up through
the wheel house. Then when he got it on top of the cabin, he put a nice,
big, shiny piece of stainless steel pipe on it. Somehow or another, in necking
it down, he necked it down from one size to another. We had the wind behind us as we started to cross the [Queen Charlotte] Sound. It's the only
open water of any consequence that you have to cross when you're going to
Alaska. It [the wind] pushed the carbon monoxide into the wheel house,
where he and I were.

All of a sudden, he passed out. The first time he did, he just [said], "God! I went to sleep at the wheel!" I was lying on his bunk. I had a real bad headache. We didn't realize it was the carbon monoxide. Anyway, the second time he went down there, then I knew something was wrong. I don't even really know to this day how I knew what it was, but I did. Instinctively, I took him out on deck. Normally I couldn't lift [him]. I just knew he had to get out of there. The kids were down below, and it was the only time that I'd ever closed the door to their sleeping quarters. But it was closed, so I couldn't call to them. I couldn't raise them at all. I got him on deck, and I knew he needed artificial respiration. [But] there wasn't room on the back deck. I rubbed his arms and slapped his face. I thought he was dead. It's the only time I've ever crossed Queen Charlotte when I didn't know what the weather was doing, and I didn't care.

Finally, I got him to say that he realized something was wrong with him. I said, "Stay where you are. I've got to get the kids." I really thought the whole time that they were probably already dead, because I knew it sinks to the bottom. I went tearing up there and opened the door. I've never been so thrilled to see three kids in my life. It's a good thing I'd closed that door. That's probably what saved them. I can have nightmares over that one, even yet—just as vivid today as the day it happened. So we anchored up and . . . went on into Prince Rupert and took care of the situation. Ran a rusty pipe up. Forget the pretty stainless steel one.

We went to Alaska the first season not knowing a soul. We got to Ketchikan with $26 in cash, three kids on board, enough groceries to last us through the summer—and we didn't know where to go. We didn't know where to start. But one old fisherman said, "Well, if it were me, I'd go to this place."

So we left Ketchikan and went out a few miles from there and put our gear in the water and fished for the first day. I don't think I've ever had a longer day in my life. The boat wasn't really finished. I had no seat to sit in. I stood there on a metal tool chest all day long, to stand at the wheel and drive. The kids were on board, and they had to be fed. Oh golly, I really got tired. I think we ended up—we had two or three fish—and that just isn't going to do it. There were no other boats there. We did this for about two days.

We had a radio on board, which was a new thing just coming into the fleet. Everybody was getting radio, so you'd have communication. But we couldn't get used to the noise of it. Finally, I said to my husband, "Why don't we turn that radio on and listen? If we can find out where the boats are, maybe there'll be some fish there." We just happened to turn the radio on at the moment [when] one of the guys, who was about thirty miles down the coast from us, came on and said, "Hey, I'm in fish right up to my ears out here, and I'm the only boat. You guys might as well come out and get them." Man, that's all we needed to hear. So we dashed back to Ketchikan. We sold $26 worth of fish, we filled up the fuel tanks again, and then we

went out. From then on it's been nothing but up. We've had good years and
better years, and bad years, and the whole thing. But essentially it's been a
decent living for us all the time.

*Helen and her husband have been Alaska fishermen ever since. Their children
fished with them for many years. In the winter the family is based in Port Angeles,
Washington. For the first five years of fishing, Keane worked in a shipyard in the
winter. Helen has regularly taken college courses.*

First year, we fished Ketchikan and that area. From there . . . we went
. . . up [north] after that. [We] mostly fished in Cross Sound and out in the
beginning of the Gulf of Alaska, out of Pelican, Sitka, that area. Our first
[boat] was a forty-foot wooden sailboat, which gave us more living space than
an ordinary troller would do. We didn't know enough about fishing to really
need a great big boat. The boat worked out beautifully for us, for seven years,
until we really learned how to fish. [Because] . . . we were a sailboat, a lot
of people resented us for a long time. They'd say, "Here comes that g.d. yacht."
For several years we heard so many times, "It's another one of those schoolteachers up here taking our money away from us."

The way we did it when we first started was just to watch the boats. There's
still a lot of that going on. Any new fishermen that come into the fleet,
they'll say to you, "Well, how will I know where the fish are?" We say, "Watch
the top boats. You'll learn quickly who's catching fish in the fleet and who
isn't. Watch those boats, and if they suddenly come up missing, you'll know
the fish have gone somewhere else. Listen to your radios."

When I had the youngsters, my husband would go up and fish alone. Then,
the day school was out, we flew up and joined him for the summer. We usually were there by the fifteenth of June. We came back sometimes even two
or three days after school started in September, because when the fish are
there, you've got to fish for them. It really wasn't any problem for the
youngsters.

[Our children] were certainly good crew. They were fishing people from
the very beginning. I . . . steered the boat . . . while my husband did the
fishing. The girls were ten and six [the first year we fished], and they took
care of the little guy, except for things like doing the laundry and that sort
of thing. They learned, too, at an early age to do the cooking for us. I did
[it] for the first few years. But when the girls were probably twelve and nine,
or something like that, it got to be a hassle, because the youngest girl ended
up just getting furious. "All I ever get to do is peel potatoes and set the table." I saw a problem developing there between the two kids; so I said, "Well,
why don't you come out and help your dad with the fishing?" Our oldest
daughter started first, cleaning the fish, and she baited the hooks. She did
all this sort of thing for her dad. Of course, the other girl by nature is really

a homemaker. This is where she really did the best job. So it worked out just fine. Ultimately, they all got involved in the fishing.

My husband would catch them, and then I would run the gear and land the fish between all three youngsters. They cleaned the fish. We'd start them out by saying, "Well, we'll give you money for the humpies. Clean the humpies and you get the money." It was a way for them to make money and also to motivate them to do fishing—which they really did.

[We were always busy on the boat.] Sometimes we spent hours running to the fishing grounds. There's always something to do on a boat—and, if nothing more, relaxation. Because when you do get into fish . . . you fish from early light to late light, so you sleep and rest for only whatever you can. If you're running for ten hours, you take turns. Somebody's on the wheel part of the time, and the rest of the crew's sleeping or eating or relaxing—whatever. But there's really never any shortage of things to do.

The first year we went up to Alaska, I went down to the store and bought a whole bunch of games and puzzles and that sort of thing. I hid them on board just for the day when the kids'd say, "There's nothing to do." Because they will on a thousand acres. So we put these things on board, knowing that we were going to probably be faced with that. It never happened. It absolutely never happened. We suffered through some real tremendous theatricals that they put on with us—with blankets and sheets and whatnot. But they entertained themselves. Of course they were below, and they never got up at the crack of dawn to go start the fishing. When they'd get up, they'd get involved in whatever down there. Sometimes maybe it'd be four, five o'clock when they were really little . . . when they'd come up and say, "Um, gee, it's rough out today." They wouldn't even be aware of . . . what was going on. If we were going to be in heavy fish, then we'd get them up there to help us. But if we weren't, why we'd just [let them be].

We made sure that they had time to play when they went ashore. We had birthday parties on board, and we had birthday parties on the docks. We just made it as much fun for them as we could. They had lots of friends in Alaska, and they had lots of friends down here in the fleet. We had two homes, it seemed like. Our kids used to say, "Gee, we get two homecomings a year." When they would get back to Alaska in the spring, all their friends from Alaska were there in the fleet. Then when they'd come home in the fall, they had the friends in Port Angeles. But we feel as much Alaskans as we do [natives of Washington]. We have as many friends in Alaska as we do right here.

When the youngsters were little . . . the fish were plentiful and close to shore, and we went into a harbor every night and anchored up. Then, after they were bigger [and] . . . we were needing more money in the industry, . . . we got our bigger boat [a fifty-foot ketch], the boat that we have now. [It is a] steel boat . . . which also sails. We started moving offshore, and we quit day-selling. We would trip-fish and stay out for ten days. That's when our income really started moving up. When our kids got big enough to where we didn't feel that . . . we were making it too hard on them to really go out to sea [we went trip-fishing].

When her children were young, Helen was afraid on the boat. She also faced criticism from people who thought that it was dangerous to take the children fishing.

Looking back through it, I really have to laugh now, because for about nine years I was scared to death of that boat. I . . . never learned to swim until I realized we were going to have a boat. Then I went with the ten-year-old kids and took swimming lessons. The first nine years we fished, I was positively petrified. I cried when we'd leave in the morning, because I was afraid we weren't going to get back at night. We were going to drown those kids. In fact, I never really got over that until we went cruising. We went to Mexico in winter when we bought this latest boat. [We] spent three or four months down there. We left Neah Bay [and had] about six hours of fine going on our way to Mexico, and then it was pure hell from there to San Francisco. God, it was awful, stormy weather. Just having to stay out there proved to me that our boat would take it—we had the capability of doing it—and that my husband wasn't completely out of his mind. We could make it. Then I got over that fear, which was a real plus for me, because society condemns you [for placing children in situations perceived as dangerous].

I got a lot of malarkey from people down here, particularly my in-laws and people who didn't really understand that we loved those kids as much as we did. They felt like we were just really taking those [kids] out in the middle of the ocean and providing them with the opportunity of drowning. We really felt pretty secure about it, or we wouldn't have done it. [But] that was a mother's fear for her kids. My husband won't buy that to this day, but it really was. Especially after what happened the first year going north with that carbon monoxide. Of course, my husband slept through it. He doesn't even know the terrible [anxiety] that I endured through that. He says, "Oh, that's just an excuse." I think he really knows better than that. But it really was a real consideration for me. As a mother, I felt very responsible for them.

[The carbon monoxide accident taught me] that I'd better learn everything possible about running that boat. [I had to] know how to handle it if he got sick, or fell overboard, or something like that. We taught the kids this, too, as they grew up. There's always a possibility of losing someone overboard. We have what we call float bags that float out behind [the boat] that some of our lines are on. We used to say, "If you fall overboard head for the float lines. Get a hold of them because that would keep you afloat."

The doctor that we were taking our kids to said, "Gee, aren't you afraid of losing them overboard?" I said, "Well, what we're really going to do is, we're going to find a calm day, we're going to put their life jackets on, and we're going to throw them overboard." He said, "Gee, isn't that kind of tough?" We said, "Well, yes. But they think the water's fun. They think it's going to be fun swimming." We know it's colder than the dickens, and that it can kill you. But if they find out that it isn't so fun ——— .

We'll none of us ever forget the day we did it. It was a beautiful, warm day. Going to Alaska that first year we had some engine troubles. So we shut down to drift while my husband repaired a heating problem. The kids—the girls especially—were jumping up and down. "Throw us in, Daddy! Make this the day!" We had told them that we were going to do this. So we put life jackets on them. Then I got in the water on a rope, so that I'd be there, and he stood on the deck and he threw them in. He threw them with such force that they would go under water and then really come popping up, because he wanted them to really be aware that the life jacket would keep them floating. Of course . . . [the weather was] so warm and beautiful, they thought it was going to be warm in that water, and it was colder than the dickens. They learned. We never lost a kid overboard—never did.

I know without a doubt, that no matter what comes up for my kids, they're better people for the discipline they had to learn on board a boat. You don't stand there and argue with a youngster and say, "You can't do that," for five minutes, because if you do, he may be over the side. So you say, "Stop it right now." They're more self-sufficient kids. They're kids, usually, that can entertain themselves. Most of them are readers, simply because there are hours and hours where there's not much else to do but [read]. My kids really consider it a plus. They'll say it right now. My son's twenty-four, and he's involved in the rat race. He says, "Oh, my God. When I get up in the morning and think I have to go in for eight hours, I wish I were in Alaska on board." They really know what it's like. What a tremendous thing it was for them. It wouldn't be for everybody. You'd have to like the water. You'd have to like the mobility; you'd have to like being here one day and someplace else the next.

Fishing Without the Children

Since the late 1970s, when their children stopped fishing with them, Helen and her husband have fished by themselves.

[When we're fishing] my husband starts the day by pulling the anchor, probably four-thirty, five o'clock in the morning, depending on whether the fish were there or not there the day before. He handles it probably till about seven o'clock, till I get up. When I get up, if he has fish on deck, then I go out and clean them. I'll get them all washed up and ready for the ice hold. Then, by that time the morning bite is probably over, and we'll have breakfast. [I'll] fix whatever we have for breakfast. Then we just continue that way throughout the day. Around noon I fix lunch, and he has his lunch. [Then] either he or I elect at that time to go down and take a nap, if conditions are so that we can, because it is a long day sometimes. Eighteen hours is not unusual. So he or I go and take our nap. Whoever is on the boat at that time and not sleeping is running the gear and doing whatever has to be done. We usually take two hours a day for a nap, except if we're in heavy fishing. You don't even need a nap when you're really in the fish. That depends somewhat on the area, too. If we're offshore where there's no other boats around or anything, why we always get a nap. But . . . if we're fishing close to shore, we sometimes have to do away with that. Because there's too many boats around, and too close to the land, and . . . it just takes two to operate the boat. That continues on until dinnertime. Once or twice a day, he'll go down and ice down the fish that are already caught and cleaned and get them in the ice. That's important to us because of quality control. I'm in there, listening to those radios. I quite often know when we should be moving to another area, because I'm listening. I'm more perceptive on picking up somebody's tone of voice, if there's fish someplace, than he is.

At dinnertime we eat, sleep, and drink practically the same as when we're at home. If it's rough weather, I'll put a roast in the oven in the morning, and it's usually ready by dinnertime. We usually try to have our dinner together. We just stop the fishing. The boat keeps going and we are essentially fishing, but we quit running the gear until we've had our dinner. Then it's just more of the same. Catch the fish, clean them, put down until dark. In Alaska, of course, it never gets dark till 11:30 or so. Most of the time, our work is done by the time we're in harbor or ready to shut down. The fish are cleaned, they're iced, and they're put away. We aim to be in bed by midnight. It doesn't always happen, but that's what we really aim [for]. That's the way it goes, day in and day out.

We have five gurdies on board, which are power reels that run our gear up and down. We have anywhere from four to six lines. Some boats have more— eight lines. We put on a heavy lead weight, sometimes fifty-pound lead weight, or thirty, or sixty, or whatever you're using. It depends on depth and lots of other things. We put the gurdy in gear [and] let the gear go down. There may be anywhere from five to twenty hooks on each line with lures on them. We have springs on our boat from the poles [so] that we can actually stand in the wheel house, and we know exactly when a fish hits the gear. After this many years, we know what kind of a fish just by watching the action on the spring. We can tell exactly what's on that line—whether it's king salmon, whether it's coho, whether it's humpie.

When we see enough action on the gear, then we go out, and we put it in gear, and we bring the gear on board. It's brought in by the mechanical gurdy.

The reels . . . are power driven, and [the lines], they're brought in. Then you take each fish off as it comes up. Take the hook out of them. Those lines can't catch a fish when they're on deck, so immediately they go back into the water. We . . . knock [the fish] in the head so that they're stunned and [then] throw them on board. Then they don't do a lot of flopping and lose their scales, which is important. They're not as pretty [if that happens]. The troll fish go to the fresh fish markets. We want really good-looking fish in the markets. The troller is known for the finest in the fresh fish market. For that reason, we get a little better price than some of the net fishery. We . . . immediately cut the gills and the blood vessels so that they bleed right away. Especially the big king salmon. The sooner you get the blood drained from them, the better off they're going to be qualitywise.

Then they're pushed forward on our boat to where I go. I'm the one that does the cleaning on board. I . . . take the insides out, take the eggs and save them . . . if they are there, because that's become a very valuable commodity in the last seven or eight years. We used to throw them over the side, by the tons. It makes us a little sick now when we realize that the market was there and we didn't realize it.

It's not done the same on every boat—it's kind of a matter of personal preference—but we wash them twice. I clean them, wash them thoroughly once, put them in a box, let them drain for about thirty minutes, then rewash them and put them down on the ice. Then he ices them in there. That makes for a better product in our opinion. It's a fairly fast process. From the time they're landed, the ideal thing would be to have them, within half an hour, down on the ice. But if you're in a lot of fish, it isn't possible. But then you cover them and keep them dampened down until things slack off and you can do it. So you do it just as quickly as you can. When the kids were aboard, it was . . . a faster process.

It's more sophisticated now, by a long ways, than it was when we first started out. That first year [in Alaska] we only had a radio and a fathometer. Now we have Loran on board that pinpoint[s] our position on the ocean. If you find a school of fish, you know exactly where those fish are. You can turn on your Loran readings and come back to that spot and probably pick up the fish again. It's become a lot more efficient. It used to be that we sold to the buyer boats, but then the fishing situation was changed so that there's more money if you deliver your own fish. [Now] a lot of it's done offshore. There's not the fish there used to be. You don't like to fish around [other] boats. You just go off farther in the ocean and find your own fish. It isn't as much fun as it used to be, because you don't get as much time to go to the beach. At least we don't take it. It is more of a business. When we first started out, we really wasted a lot of time and a lot of money. But I'll never regret it, because it gave the kids time to play on the beach.

Helen and her husband have a specific division of labor when they work together on the boat. In such a close working situation, compromises have to be made by both of them.

[So], I clean the fish and he runs the gear. I don't pull the anchor. Some women do, but I never had to. As far as running the boat, of course, I can handle anything that comes up. I don't do the mechanical work. Some gals are pretty sharp on that, but I'm not interested in that. That's an area where I'm sure I would get in his hair if I did. That's his thing. I don't know that he *likes* to, but then he considers that his area, and that's fine.

I find it kind of interesting, because he doesn't even really want me back there pulling gear when he's back there. It's fine when he goes down for his nap, but we never work in the pit together. That's been a kind of irritating thing to me, because I've wanted to. I like the competition. I'd like to say to him, "Hey, I'll bet you today." Because a lot of wives [and] husbands do this sort of thing. But he's not a competitive fellow at all, and he just doesn't want it that way. It's an area I leave alone, because I can see where there could be problems. And, hell, we don't need to create problems for one another. I'm a real strong, opinionated woman. I'd have to say he's probably given a lot more than I have in the marriage. I am, by nature, explosive and volatile, and he's very subdued. He's a lot of fun—don't get me wrong—but he's not forceful at all. And, thank God! We'd have probably killed each other a long time ago had he been. There's conflict sometimes. He doesn't run the boat the way I think he should, and I don't run it the way [he thinks]. When you're in a small space like that, you're just going to rub each other wrong sometimes. But then, what the hell. That's what life's all about.

Basically it's worked out very, very well. For me it's been thirty years of being right with this man, because I worked with him in the tavern, and I worked with him on anything that he's ever done. It's the way I wanted it. It's why I wanted to go to Alaska, why I wanted to take the kids. Because I didn't get married to be alone. I got married to share my life with my husband. I think it's really neat that we found something that we could be successful at that we both liked. It's a big plus for him, too, to have a wife that's willing to do this. It's a plus for me that he would put up with me for the nine years I was scared. I know . . . there are men in the fleet who would no more let their wives go on board than fly to the moon. I do know men that would like to have their women down there painting, but would not have them on board while they're fishing. I don't know [why]. I think it's because they feel threatened by it. Or maybe they'd have to take it a little easier, not fish as hard if [their wives were on board]. I know there are men with their families feel this same way. They say it's too tough on the kids. We haven't found it that way at all. That's just their way of explaining why they don't want it. It's not necessarily true all the time. [At] least [that] hasn't been our experience in Alaska.

I consider myself—on paper I'm not—a member of the Washington Trollers, the Alaska Trollers, [and] a member of the Halibut Producers Cooperative [HPC]. [My husband's] chairman of the board of the HPC. I just don't consider myself not a part of the fishing fleet. I know that the fellows talk to me and value what I have to say. They know that I keep up on what I can, and that I'm interested. But as far as owning a share in it, and whatnot, I own the half that's mine by law [*Washington is a community property state—ED.*] and knowing that's it.

Plans for the Future

Helen and her husband hope to continue fishing in Alaska during the summers and to take their boat cruising abroad during the winters. They have already sailed to Mexico, and they took a two-year trip to the South Pacific with their two younger children in 1969. They are concerned that the increased regulation of the salmon fisheries may keep them from realizing their plans.

We're beginning to wonder how long they're going to allow us to [fish]. The regulations that are being promulgated are just incredible to us. Traditionally trollers have moved up and down the coast with the schools of fish in the ocean. Now . . . they're saying, "You can only fish Washington, or Oregon, or Alaska." This is ultimately what they're after, no question about it. The troller has followed the fish. If he didn't find it in Alaska, he came down to Washington, because with the radio situation now, you know what's going on. [Now] because there's limited entry, that's really a thing of the past. So far, we still have Alaska permits and . . . Washington licenses, but they're fast trying to change that. That's why it's really becoming such a restricted fishery. Part of the reason [for my son not staying in fishing] I'm sure is that—well, how *can* you? It's limited entry. The only way he could get into fishing is to buy a license *and* a boat. For the young people, it's really a bad deal. In fact, it infuriates me—not only me, but many, many people—because of this very thing that's happened. Here's a youngster who's put aboard, raised *entirely* in this situation; yet the only way he could get into it involves at least $100,000. That's kind of a sad commentary.

I understand why the fisheries department wants to do this, because we are uncontrolled in comparison to the net fishery. But always the thing that bothers me as a troller, and bothers every troller, is that none of us actually feels that a fish run could be wiped out with hook and line, because they have to bite our hooks. It isn't like a net. You could put a net out someplace and catch every fish. But we feel put upon and pushed and pressured, simply because [a] fish has to bite our hook before we even get a chance at him. Yet they say that we are the baddies in the industry. We're killing the runs. Well, it's a crock as far as I'm concerned.

We find ourselves wondering if . . . [fishing's] going to last as long as we need it to. We're in our midfifties now. We conceivably have ten more years of fishing time. I can't imagine anything worse than an eight-hour day. We've often speculated, my husband and I, what we would do if . . . there [were] no more trolling. The last thing we'd do, I'm sure, would be go back to working eight hours a day at something. We'd probably grab our boat and go sailing the South Pacific. It doesn't take all that much money. So, we'll see. Hopefully we can fish Alaska [and] cruise the rest of the year. Be a neat way to wind it up, I think.

Lois Engelson

Lois Engelson

Lois Engelson, born Eloise Fields in the 1920s, was raised in Everett, Washington. She says, "My family, we've all been seafaring peo-ple." Her father operated a tugboat, and some of her neighbors were fishermen.

Life Before Commercial Fishing

[My mom and dad] were married on my mother's fifteenth birthday. My father and another fellow used to have a tugboat, and my mother . . . was on it [with them]. I was four days from being born on that darn tug. My mother had to get out of there. So, I think that's where I got rid of my seasickness.

Where we lived, in Everett, was a fishing family down the street. The woman prepared the meals, and usually a man went out and then came back in the evening, or just maybe overnight. Not trip-fishing. [The women] just didn't do things like [fishing]. As [kids], Sam . . . and I took his father's rowboat out when [we weren't] supposed to—silly things like that.

As a young girl, Lois demonstrated interest in learning mechanical work. At seventeen she joined the army to learn more about it.

My father said [that] ever since I was a little kid, Momma [would] put me in a clean dress [and] if he'd be under the car, I was under there right with him. So, ever since I've been two years old, that's been predominant. When I was thirteen, I wanted to be a diesel engineer. My father said, "No girl in our family is going to do man's work."

[I] went in [the army] when it was WAAC . . . in [1942], when I was seventeen; when it turned over to WAC, I got out. I was in there eleven months and two days or so. I started working in mechanical work . . . in Daytona Beach, Florida. [I joined the army] for the going overseas . . . [and] for the education. I took a test for mechanical aptitude and [for] radio; those were my two high ones. . . . I had a choice of which one I wanted to go to. [I chose mechanics]. I don't remember where I was at all the different times, [but] I ended up at Camp Roberts, California, in the desert. [The WAAC] just kind of ended up to be a null, actually, [when] it turned over to WAC. [So I got out.]

A few years later Lois married her first husband, Tony Cassell. He was a logger, and Lois worked for Stevens Motor Supply in Forks, Washington. They had been married for four years when her husband died in 1951. She continued to live on the Hoh River for a number of years after his death, working in Forks, fishing, and doing odd jobs to support herself.

I used to run nets . . . down the Hoh River when I lived down there [for] five years. Part of my living was gotten off that. I [ran] what they call a drift net. That's two poles with a net on one pole, and you push it out and then drift down the river. Indians had fish nets in the river, crisscross, crisscrossed, all the way up, ever since I can remember—clear back in the forties. I never had no trouble. I lived on the Hoh; I could always catch fish.

I never had beef for over five years—bear meat, deer meat, elk meat. I knew how to cure it. I put it up and everything. A bunch of guys that were working on the highway down there asked me about fixing their lunches, because they couldn't get them fixed in Forks. This one fellow . . . is supposed to have [had] real bad ulcers. Every time he got a sandwich, he said that if a truck drove over it . . . it'd be nothing but one greasy skid mark. So I said, "Sure, I'll fix your sandwiches." I was doing the sandwiches for $1.00 [per] lunch. I was baking all my own bread . . . and I was using bear meat in their lunch. Poor guys . . . didn't know the difference. They complimented me on the best meat they ever ate in their lives. One of the guys knew that it was either bear, or deer, or something, because that's all they ever ate. They never ate beef.

Then I went to work in the woods. I was whistle punk [for a logging operation for one and a half years]. I had to have my hair braided and up underneath the hat so that the other ones didn't know I was a girl. My checks were made out with only my two initials and my last name—not in my [first] name. The guy that was running the show knew me, and he knew that I needed to make a living because Stevens Motor Supply had sold out . . . and that left me out of a job. In those days $2.05 an hour was a pretty good wage. It was Francis Jarnigan's outfit that I worked for. Francis Jarnigan, he found out I was a woman. But I'd already been there for a while, so he says that as long as I know my job that he didn't care.

While still working at Stevens Motor Supply, Lois met Clyde Fasola, the man who later became her second husband.

I [first] met him when I was [still] working at Stevens Motor Supply [after Tony died], and he knew I was living by myself down on the Hoh River. He . . . sent a fellow down there in the wee hours of the morning. I [lived] on a twenty-foot bank up from where [the fellow had] parked. I [got] my Swedish Luger down. I went to the door . . . and [the fellow] said he was having trouble with the gas line leaking—sucking air or something. I told him, "Well, okay. I'll get some soap." I figured I'd soap it up, because that's all you do is put some soap on the line. I was talking through the door. I hadn't opened it yet. I got a bar of soap . . . and opened the door, and he stuck his foot in the door. He says, "There's nothing wrong with my car. I came to see you." Holy catfish! I pulled that Luger out and he could see it. He had a nice new suit of clothes on, and it had been raining. He ran down that bank, then fell head over heels. I laughed all the way while he went to the bottom. Day or so later . . . I saw Clyde, and Clyde says, "Boy, you just about ruined my friend." I says, "What do you mean?" And he says that he was the one that'd sent him down there to see me. He just wanted to see if I was a woman of the world, I guess. So I didn't marry him then. I went back to Philadelphia to school [and then to California].

Lois went to Pennsylvania to study accounting at Upper Darby Business College to become a certified public accountant. After studying there for some months, she discovered that the license she would receive would not be valid on the West Coast. She left the program and went to California.

I worked for Lindsey Lumber Company in Fortuna, California . . . as a full-charge bookkeeper. I don't remember just how I got that job. I had a boat . . . down there. There was a boat that'd got a hole in it, and somebody told me that I could have it for $125. I didn't know anything about it, and it looked awful good to me, so I bought it. They brought it on a trailer and parked it in back of the office where I worked. Then, finally, I took it home—had somebody drag it up there. The wood . . . looked pretty rotten to me, so I started tearing the boards off of it. [Then] a fellow came up to me and said, "What are you doing?" I says, "Well, I've got to fix this boat. It's got a hole in it." Well [it turned out] it was upland cedar, and it's naturally soft. So I was tearing apart stuff that didn't need to be pulled apart. So then he started helping me on that. I never did put [it] on the water [but] I turned around and sold it for $500, and that was my first experience with a boat. I used to fish . . . across Humboldt Bar with a boat, but that was just sportfishing; I took another lady friend.

[When I was working in Fortuna] I used to come up here every winter, because I had my girl friend in Forks. I stopped into Babe's Tavern [in Port Angeles] to have a beer and see if there was somebody around that I knew, because I knew a lot of people here [in Port Angeles]. And [I saw] Clyde, and that was it. I never went back to California.

Learning to Fish

Lois married Clyde Fasola in 1955 and they settled in Port Angeles. She soon found an opportunity to buy a boat. This marked the beginning of her career in commercial fishing.

I got my boat in 1955, 1956—something like that. There's a woman, Mary MacDonald, here [in Port Angeles]. Her sister owned a rest home. There was a fisherman that was there, and he owed her some money, so he gave her the boat. Mary . . . asked me if I'd like to have a troller . . . it was thirty-two foot and a double-ender. I says, "Boy, I would, because I'm always down there on boats." I always loved boats. So I come home and asked my husband if I could get a boat. [He said], "Yeah, we can get a boat and have it for job security." Well, it ended up that security went to me. When we divorced [in 1961], he took the house and the car, and I took the pickup and the boat. [Clyde and I are still] friends. Bless him, he's Portuguese. He keeps saying to me, "Well, I wonder . . . what we'd have today if we'd stayed married, if you hadn't rejected our contract." *Nueva Esperanza* is what I named [that boat]. The name, translated [into] English, is "New Hope."

I went out first on lingcod fishing by myself, and then when [Clyde] was off, because he worked during the week, he went. [We were] fishing forty-five minutes from here [Port Angeles]. That was drifting and trolling—what they call "handline." You'd have just the gurdy down. Well, I was always down on the boat, working on the boat, and then I'd come back and fix supper for Clyde. I always wanted to go out and . . . go fishing. Well, of course, he had to work. He's a logging-truck driver. The big decision came up one time, and I said, "Okay, I'm going to go fishing." So he said, "Okay." So I went . . . drift fishing. I was going to do like the rest of the fellows. [At] first all I had was a compass and the gurdies. I didn't have a DF [direction finder] or nothing.

Well, I come home and bring my fish check home. I was real proud of it, because my fish check was bigger than . . . his paycheck. Finally, one day, I was [away fishing, and I was] anchored. I was busy working on something, and I felt something against the boat. I came up there, and there was Clyde. He had hired somebody to take him out there. He figured I was shacked up with somebody. His evil mind! And he says, "Where is he?" And I says, "Where is who?" And he says, "Well, you can't be running around here all by yourself doing all that fishing and bring home that much money." But he found out I was. It was a weekend; so he stayed that next day. So then he says, "I'd rather you not—we'll just go fishing on weekends when I can get off work." He was a jealous, crazy old man—he ever was jealous of me. We came home, and . . . I wasn't going to wait till next weekend. I said, "There's good tides right now, and I want to hit those tides." If you're going to fish commercially, you have to be able to [go when] the tide is right. And I was coming in with pretty fish. Everybody asked me where I got them, and I'd tell them, "Out there." I'd just bubble over . . . and be so proud.

Clyde and I got divorced. That was one divorce. It cost $10 for the attorney—we had everything all agreed—and $27.50 for court costs. That was it. He said to me, "That boat's more important to you than I am." And I said, "Yes." [Fishermen's wives don't feel that fishing is more important to their husbands than their families are], because that's their grubstake. The husband is the one that brings home the wallet.

Lois fished alone on the Nueva Esperanza, *trolling off the Washington coast and gradually expanding her activities to Alaska. She learned fishing and boating skills through practice and through trial and error.*

[It] would have been really my desire . . . to be able to work on a boat for at least three or four years, or seasons, and then go own a boat. That's the way I would have done it, if I could have. [At first] I didn't even know the difference between the fish. I got the book out and . . . I looked at that fish. Well, does it have them kind of scales? Is it like that? Now, the instant I see a fish [I know].

I didn't know how to sharpen hooks or nothing. But there was this fellow here in town, I called him "Dad Barber." He was an older fellow, and he kind of took a liking to me because I wanted to fish. There at the dock . . . he started showing me different gear—like, if I had number 5 plug, using them if I've used number 6 hook. Same with the spoons for fishing—brass on a dull day, and bronze or chrome on a bright day. Speeds—if you catch your fish when you're putting your gear down, you're going too fast. If . . . fish get on your gear when you're pulling it in—you have several lines on a gurdy—you're going too slow.

[Early on, when] I was out there and had no equipment except the compass—and I got a radio—I would let my leadline down to 100 fathoms. Then I could kind of work it in from there to get into wherever I was going. I always kept track of the direction of the wind and how strong it was, not really knowing what I was doing. But I knew I was doing something that was going to benefit me. I got a book given to me in Seattle [written by] Bowditch, which is any seaman's Bible.[1] In back it's got all your tables . . . all those kinds of goodies for figuring your height . . . wind-driven currents, and all this kind of stuff. I used to sit and study that by the hours.

Being a woman, when I first started fishing, I was never going to tell anybody I was lost at sea. I was not going to get lost. Well, as all fishermen listen on their radios, if you mention something about catching a fish, they can take a direction finder—and zoom, they got you. And they run right out to you. I figured if I ever got lost, I'd just call a friend of mine and tell him, "I've just got into a whole school of fish." Then I'd never be lost no more. I've never had to use that, but I always figured that was my ace in the hole. I always took pictures when I come out of a place I didn't know, so I could look back. I mark them. That's one advantage with a Polaroid camera.

[Like other fishermen, at first I used radio conversation to help me find fish. I was] listening to other people talk in codes [on the CB radio]. In the morning [a] fellow says, "Well, it's mighty slow." In the afternoon, it's "pretty slow." Well, you . . . know that he's got more fish than what he had. Say, if a [fellow] has some pretty good fish [and] he's someplace else, and he's talking to his buddy, and his buddy's [boat is] around me [the fellow will] say, "See you on the next tide." Well, I know who he's talking to, so I've got my binoculars on nearby vessels looking for [this buddy], and when he starts picking up his gear, I pick up mine, too. Just s-l-o-w-l-y gather away from the rest of whoever's fishing.

---

1. Defense Mapping Agency, *American Practical Navigator: An Epitome of Navigation*, originally by Nathaniel Bowditch, 1773–1838 (Hydrographic/Topographic Center, 1984).

[One time] my lines tightened on the capstan [so] that I couldn't get it undone without having to cut that nylon rope. And I didn't want to cut that, because that was $308 worth of raw line. So a friend of mine rowed his dinghy over . . . and he says, "What's the matter, Lois?" because he saw I was having a problem. So I told him [and] he says, "Oh my God! I'll . . . help you." So . . . they pulled their gear—he had a deckhand with him—and tied [the line] off from the stern of the boat. He came on board and took that off. Boy, between the two of us, we finally got it. So he helped me get it undone . . . then he showed how to properly tie off. Of course, on [my] last boat [I] didn't have to bother about that because I had a big anchor winch in the front.

My most loveliest thing [was], I used to practice docking. I'd go out to that log boom and I'd practice and I'd practice. I'd have a certain ring I had to meet, and pick it up, and tie my lines to it. I'd practice day after day, especially in the wind, because I had to learn. A couple of years later, when I was in La Push getting some fuel, there was a gal that had a boat. It wasn't a *real* fishing boat; it was one that somebody stuck some poles on. She came up to the dock to get some fuel, and she didn't know how. I told her to put her bow against the pilings there . . . so she could swing her stern over. She says, "Well, I don't know how." I says, "Well, if you don't know how . . . you have no business being out here." And that was only my second year.

The first experience I had going up the Columbia River was rather unique. I knew [the] buoy system and so forth, but I didn't really realize how vicious the freshets can be in the springtime. I knew that . . . whenever you go over a bar . . . you always used to come on a high slack. A lot of this information came from my father. The high slack was about five minutes after midnight . . . so . . . it was dark. I could see red lights; I could see green lights. I had them all sorted out in my mind which ones I should follow. And then I looked and I saw lights coming from every place. I saw red lights, and then they was turning green, and . . . I thought, "My gosh! What's going on here?" So I called a friend of mine . . . Frank. He says, "I forgot to tell you. The dredge is right down in that area. Possibly this time of the night what you're seeing is some ships coming out." He says, "Just hug real close to those first red buoys there coming up." So I did, and . . . sure enough, there was five ships. I kept looking at my chart, and I kept looking up [at the lights] till finally I got them all straightened out. And I got up there, and then I tied [alongside] Frank's brother's boat. The next morning when I woke up, the velocity of the water was absolutely appalling. It's the first time I'd been up the river . . . but I have charted the area and I kind of studied the chart out. When you take and read something, it's like looking at this picture. Seeing the actual location, you get a different perspective. That's what happened here.

The thing that attracted me to fishing . . . was looking for the fish and [being] able to try to find the fish by myself. It was a challenge. [Fishing] is hard; it's not difficult. There's no difficulty in working if you keep organized. I didn't even give it a thought [that there weren't any other women fishing].

Fishing up in Alaska on the inside—not on the Fairweather Grounds, but when I was fishing in close, . . . you follow a fathom curve. The [bottom] is charted in fathoms. In Alaska that curve is usually right on a shelf. So if you're fishing and the boats are fishing [one] way, you don't turn around and come back fishing [that] way; you make your circle. I knew this before I went up, because some other people had told me. So when I started fishing . . . I fished with the rest of them; I didn't try to backtrack or nothing.

There was some people up there—they didn't particularly want me to fish. One guy, his name was Frank, I forget what boat he had. Right around [there it's] forty fathoms. Then there's a spot right in [there that] comes up to thirteen fathoms. If you've got your gear out, you can imagine what would happen. It [would] just take all your gear right off. So, before I would ever take anybody's word for it, because there are uncharted spots, I'd have the gear just in the water. Look like I had all my gear out, but I'd just have it in the water, no more than, say, maybe five fathoms down. I'd follow, because this guy [Frank] told me I could follow him. Well, he led me right over this thirteen-fathoms spot, and on my recorder it just came right up like this and came right down again. I kept right on going and he says, "How'd you do?" I says, "That was pretty good. I got a nice big salmon." He says, "Did you lose any gear?" I says, "No. Was I supposed to?" I had been warned about these tricks. Well, he never pulled that trick on me again.

Commercial Fishing

As Lois gained experience, she expanded her fishing activities to include both salmon and albacore and began ranging from Alaska to California during the regular commercial seasons. During the early 1960s Lois married Robert Townsend, also a commercial fisherman. Some seasons they fished together on Robert's boat, the Pocahontas. Other seasons, while Robert and his son fished the Pocahontas, Lois fished alone using a leased boat, the Nora K. Lois's marriage to Robert did not last. She says "My third husband was Robert L. Townsend, whom I also divorced. He went fishing with me, but he couldn't handle how involved I was in fishing."

During the 1960s, Lois established a fishing routine that she followed until her retirement in 1974. In 1973 Lois married her fourth husband, Keith Engelson, who fished with her during that last year.

[I used] to leave here around the last week in March and go to Alaska. [I'd] fish up there until about the second week in July, then either come down the Inside Passage or the outside coast, not fishing down. You come right on down with the fish, either delivering Port Angeles . . . or Astoria, or go down to Charlestown. Then I'd go on down to California and usually go into Frisco and ice up. That's terrible ice, anyhow, because they bring in the trucks and the blocks . . . are half melting while you're standing there looking at it. They put it in the chipper [and] blow it in the boat. Then [I'd] go out to Farallons. [That's] off of San Francisco Bay, twenty-some miles. I fished for albacore down there . . . till the World Series come on [in] October. [It] seemed like . . . three or four years in a row I watched the World Series off the San Francisco Bay. There's a little bay up in there . . . [where] I got very good reception on my TV. When the World Series time was come on, it was time for me to go home.

Down there you didn't have to unload your fish. Soon as you come into the dock . . . there was about seven buyers at Moss Landing. But there was one place I sold to—the guy always had . . . a Scottish tam on—I was just satisfied with his weights. The fishermen there brought in bonito, [which] is stronger tuna fish. The farmers would bring cases of bell peppers, of onions, of artichokes—you name it . . . for exchange. You could get just about as much of any kind of garden product you wanted. As soon as you got there— man, you had two guys jump down in the hold and start pewing your fish out of the boat and everything—wash it down—I mean, completely. You didn't have to pay them, even. They told me if I came in with over 2,000 pounds each time, wouldn't matter what I had, I got a free meal at their restaurant . . . a beautiful restaurant in Monterey. It's where you put on a dress and go to eat. I found that out in a hurry; so I had to buy me a dress. They had good food and dancing, and I love to dance.

I fished Southeastern [Alaska] from Cape Spencer to Cape Fairweather— on the Fairweather Grounds, we call it. Then I started fishing farther down Alaska's southern coast, because the fishing got a little poor up there in the Fairweather Grounds. I fished . . . out of Craig or Pelican. In Pelican . . . when you need parts, the best thing to do is to get on the plane and come back down to the states, get your parts and come back [Alaska became a state in 1959—ED.] You save time and money, because otherwise you'd be twelve days waiting for parts. So whenever I'd go after parts—I knew quite a few of the townspeople—I got some of the strangest requests, [for] paint rollers and things like that. I'd bring back everything I could. One time I brought back . . . a flat of strawberries. I thought I wasn't going to make it to Pelican with them, but I did finally. There was only one bar in Pelican . . . and at that time you could go in there and get a good steak dinner . . . for $2.75. Down in Craig I met a gal . . . her name was Joyce; she's an Indian gal. She had a thirty-six-foot troller and she fished . . . with Margaret. Margaret . . . must have been four times as big as Joyce was. Joyce . . . started this restaurant up there called "The Highliner." Whenever a bunch of fishermen came in, did she like the group? We'd have a big dinner and close the doors and have a party, you betcha!

I liked all [the fishing]. Each [kind] had their [special] desires for me that I looked forward to. I looked forward to the spring to fish the salmon; then I looked forward to summer to go after albacore. I'm not considered a halibut fisherman, because you rig up different for halibut; I caught them incidentally. [They're] difficult to handle, not to fish; sometimes you can't get rid of them. Halibuts are in halibut beds . . . on a sandy slope. There's a spot before you get to Swiftsure, . . . it's in forty-four fathoms. That's all halibut bed. That's another thing—you've got to know your grounds, what kind of area you're in, whether you get an upwelling for feed that your fish are going to be in. King salmon—if they're over thirty pounds, they're called a "tyee"; they're also called "blackmouth." If you get them deep down, usually they stay in the same area. If you get one that's . . . closer to the surface, . . . five fathoms or something like that, and you're in thirty fathoms, he's a traveler. It's moving, see. [You] catch the salmon that way. Silvers run in schools. And there, usually, you start fishing them about—well, I used to start, I think it was June 15 when the season opened. But when I first started fishing, it was no open season, and you start fishing in—um—gosh, they got the seasons so fouled up now, I couldn't keep track. But I used to fish them until they quit running or the water cooled down.

[For] albacore you run out to sea to a thousand-fathom curve. Then you start following your thousand-fathom curve with Loran readings, because that gives upwelling and your feed for albacore, which is tuna, but it's called the "longfin." Usually I start throwing the gear in at fifty-four degrees—that's on the cold-water edge or upper—and using green and yellow jigs.

When Keith and I got married, which was in '73, I took him out tuna fishing. We started out salmon fishing beginning of the season, and we went down to the Columbia River. [We heard] there were some tuna off the west coast of Vancouver Island; we ran up from the Columbia River. We were about 140 miles off of Cape Cook. I was running up at night and then in the daytime. So we were still running the next day—and all of a sudden I had tuna on the gear. There was some other fellows been kind of traveling up behind us that night. [I'd been] kind of talking to them. I didn't know who they were, but I told them [about the tuna]. We'd been talking so long that they knew where I was, I knew where they were. So if anybody else listened they wouldn't have known—but I told them. I says, "Just keep right on coming." I says, "They're just—we've hit them!" And we went into a circle and pulled 180 albacore out of the circle! They was averaging about sixteen pounds. Being as we run so far, I was trying to watch the fuel, because we were so far out. Then another friend that Keith knows, he had his fifty-foot boat coming out from Neah Bay. Later, between the garden hoses aboard each vessel, we fueled up, putting . . . about a hundred gallons of fuel oil in the boat because we had to get clear back to Neah Bay. We caught fish almost all the way to Swiftsure on a . . . straight tack. We had fish all over the deck and everyplace. [I wasn't usually that lucky]. I've had some good trips, though.

The most fish I pulled by myself was 315 silvers in one day, and that meant icing down clear till three o'clock in the morning. That was when yours truly was so tired [she] couldn't eat. But I had leaching tanks, so when I was catching a whole bunch, I put them in the tanks. There was fresh sea water running through them all the time. [There were] two sides to them—one for cleaned and one for not-cleaned fish. When you get a little, little tiny bit of slack, you just always keep your knives sharp, always keep them handy, and start cleaning fish real fast. That's something else I enjoyed; I love to clean fish. Oh boy, you betcha. I can clean fish real fast; I was cleaning them about one a minute.

When a troller catches fish, that's your ultimate delicacy of fish, because . . .troll fish are not wounded. They have to be [cleaned right away], especially on silvers; if they are not cleaned within an hour or so, they start belly burn. A belly burn is when the bones . . . come loose on the inside, because the silvers have . . . high acid. That high quality is very important to anybody who trolls. When you troll, you're only going after king salmon and silvers—your principal fish. Your humpies are—well, they'll bite, they're soft mouth. But they're what they call "pinks" in the can. They're a cheaper food product. Then you got your reds—your sockeye—and it's extremely rare to catch them on a hook. Although I did learn from the Canadians [one time] how to hook [a sockeye]. The hook was never in the mouth on any one I caught. See, they're a gill-feeder, so . . . the lure is—I'm not going to tell all the secrets, because it's a real good one.

[I didn't fish in a fleet]. A fleet is a tightly knitted group of friends that try to eat each other's throat. That's about the way it goes. I fish with a group, no particular group, because when I first started fishing . . . you had to kind of prove yourself, if you could handle the boat.

[I didn't have deckhands] until later. I had my girl friend's son Steve with me . . . back in '64, '65, somewheres around there. He'd fly up to Alaska when school was out, before he went to college. He was quite a reader. I didn't have to worry about him with alcohol, tobacco, or nothing; he just wasn't that kind of kid. But I had to hide all the newspapers and magazines, because he'd sit on the back of the freezer. Instead of watching the gear, he'd have his nose in a damn book or something. I could tell him, "Now, Steve, I'm going to depend on you. I'd like to lay down and take a nap." I could depend on him only that time. If I was up, and he knew I was up, forget it. But he was good—dependable to have on the boat.

[After that], two years in a row, I went fishing with a broken leg. I had some friends of mine, [their] daughter Tammi [was] with me . . . when she was twelve and thirteen. Then when she became fourteen, she got thinking about boys, and that was the end of that. We started out one day and she got kind of cocky, . . . so I called her mother through the Astoria marine operator and told her to meet me in Neah Bay. "Hey, [you] got a daughter coming back." Tammi was pretty good, except she liked Ramen noodles for breakfast.

Then I had another gal down in California one year; she liked to surf. She upchucked so bad . . . so I just had her eating eggs and all my crackers and wafers and stuff like that. She . . . got over her seasickness. We were out there for nine days and came in, and she averaged out $52 a day. That was the most money she's ever made in her life, . . . and I'm only paying her 10 percent.

Lois has often been in dangerous situations while fishing; she has survived storms and tidal waves. But she has always been confident in her ability to take care of herself and her boat.

I've always had confidence. I never had a fear of drowning. I've been in some pretty bad storms too; I've hit some dandies [but] I always used my cool. I was never afraid of the sea. And I was never seasick, but sometimes I was sick of the sea. Sometimes I wondered if I was ever going to come out of it. Well, so what? I intend to be buried at sea, anyhow, so what's the difference whether I get there now or then? If it's my turn to go, it's that time.

For instance, one storm off Newport, . . . it was in the early 70s when they lost eleven boats. [I was in] *Opal L* . . . a forty-two-footer. I was sixty-four miles offshore, . . . alone on the boat with the dog. Animals are smarter than people. He knew whenever we was getting into rough weather. Because he'd never drink water and he'd eat very little, then [I'd know] something was up. He was my little tattletale there. [This storm] laid the boat clear over forty-five degrees on its side. It took the scuppers; I mean the water was clear over the starboard rail. The dog was sitting there and it threw him clear up over the top of the DF [direction finder] and on the deck. I don't know how fast I was going, but I swear to God, I must have been going fifty miles an hour. Got on one of those waves and [the boat] just . . . surfed and broke the pole. And I was standing on the side of the bulkhead trying to talk to my girl friend that was out on the tuna grounds; she said it was perfectly calm out there. The wind was blowing over seventy knots at Newport, so I couldn't go back in. I started edging out to sea, and I came off of one great big wave, and I came down just as flat as this table . . . just like somebody'd chalked off the line. And oh, what a mess! Everybody knew Lois had two of everything or more. I had six little radios, two big phones—only one radar, though—and I had two banks of those great big batteries. These batteries were all [held] in with two-by-fours, and that [boat] heeled over so hard that it broke them loose. Then I hadn't locked all my drawers [so there were] papers . . . and books . . . all over everything. Pots and pans was flying every place. I had a mess to clean up, but I got out there and it was worth it; I got a good load of tuna.

[Then] when that tidal wave came [on March 27, 1964], I was at La Push. I was tied to another friend—they didn't have docks in there. I had bought brand new . . . lines. I heard it go "schnap!" I came flying up—and, of course, me in my nightgown. I looked out there and I thought, "Well, damn it! They're brand new lines." Then I looked and I saw that they didn't break. The boats were swinging around, and I couldn't figure [it] out. I started the engine first, and I laid on the horn. Glen came up, and I said . . ., "Jesus! Look! We can see right down the street!" Eleven feet up here, and all of a sudden you can see right down the whole fish market. All you could see was the little top part of the sign sticking up around the edge. Glen says, "Oh my God! Get untied!" He had to get his engine started. "When this thing goes," he says, "James Island—we gotta watch out for James Island"—because you have to make that curve to get out. The first thing I could think of was—call the Coast Guard. So I called the Coast Guard and woke them up. They said they'd heard a warning of a tidal wave, but they didn't know it was going to affect us clear down here. I says, "Why didn't you tell us? I listen to the Coast Guard broadcast all the time." We were going up the river [backwards] because of the force. [Our engines] were going full bore ahead [but we were] . . . being pushed back. Then she started to break loose. My God! I'm telling you, I never went so fast in that boat in my life.

[James] Island just "el-zippoed," and [we] went right over the top of everything real fine, because there was such a volume of water. For three days we had to sit out to sea. Of course it was nothing but just raise and fall like that. That was kind of a scary thing. Couldn't really fish, because I didn't really know what was going on.

Lois took care of all aspects of her boat maintenance. From time to time she encountered people who were surprised to see a woman participating in traditionally male activities.

I worked on my own radar . . . and the painting. That is one of the least favorite—painting. I liked working on the engine, of course. I put a new engine in the fifth boat in '65. Another thing, too, in fishing [is] noise from a prop. Even if you're in fish clear up to your gunnels there, you'll not catch them if you've got a whistle . . . in a prop; especially in tuna fishing. Many tuna fishermen change wheels, or props, when they're going tuna fishing. [If] they get a whistle in the water . . . it echoes. [It scares the fish.] You know how these whales pick up their sounds from miles and miles; well, that does the same thing. If you ever want to hear anything, just get down in the boat. I've done that when I first started fishing. My father'd finally conceded to me fishing, and [what] he said, being as I had no other way of telling, was to put my ear against the hull . . . [to] be below the water, and listen. And you could hear a ship—"tschk, tschk, tschk, tschk." You could hear that for m-i-l-e-s away.

I overhauled my engine one time up [in Canada]. This young kid . . . was watching me all the time. So I got everything apart and [was] doing what I was doing, cleaning all the garbage off the valves because they were pretty badly burned. Pretty soon the kid's father come down and tell him it was time for him to go to bed. And [the father] says, "I want to see if it'll run after she's got through ruining it." Of course, when I got it going about ten o'clock . . . the kid was so happy. [He said] "Wait till I go home and tell my mother! Wait till I go home and tell my mother!" I mean, everybody's been real good to me.

They didn't used to have women's provisions in the different places. When I delivered fish in Newport, coming in with a load of albacore, they were just starting to fix the women's restrooms [and] showers. So some man said, "But you can go in the men's." Okay. Well the men's got four or five showers. So I waited, and I waited, and I waited—and there's [still] somebody in [there]. Their door wasn't locked, so finally I got tired of waiting. I looked in and I hear the shower going—fine and dandy—so I went in and took my clothes off in a hurry and I jumped in the shower. Pretty soon this fellow comes out of another shower stall. He says, "By God! Is there a woman in here?" I says, "Yes, I'm taking a shower. I waited long enough." He says, "My God! My wife will kill me!" But, I mean—such things as that—some funny things happened.

Work in the Coast Guard Auxiliary

At about the same time as she started fishing commercially, Lois joined the Coast Guard Auxiliary. As part of the Auxiliary, she participated in numerous search-and-rescue operations. In 1961 she towed a total of eleven boats into port and earned the name "Tugboat Annie." She has also taught classes in marine skills and has helped many people informally with both mechanical and navigational problems.

I think the majority of women started in [the Auxiliary] . . . in the early sixties, because I had a hell of a battle clear down into '55 getting in, getting qualified. You don't just join the Auxiliary. You go in; then you have to pass all these exams, such as for piloting and navigation. At that time there was twelve different exams; you had to pass all of them, send in to the district [to be] approved and so forth, and then you became basically qualified. Some of the fellows that run this were a close-knit bunch. They didn't want any women in there, but they couldn't get them out. So they tried to blackmail

me—like taking the charts—because I'd passed everything else but the piloting problem itself. You had a time limit to go through all these tests. But
Dr. Miller—he was in the district at the time [and a] good friend of mine—
[so] I got hold of him. We went into the district [together] and got my test
taken. And I passed it 97 percent. After I got it—was basically qualified—
these two jokers dropped out [because I joined.] They both had other women
on the side, so their excuse when they left home [was], "I'm going to the
Auxiliary meeting." Then they'd go to the Auxiliary meeting and . . . they'd
meet the girl friends afterwards.

The Coast Guard Auxiliary is the right arm of the Coast Guard. Being
qualified, you can be called at any time for patrol duty, or for search-and-
rescue and things like that. Last year we were available to man the radio
station . . . because they were very shorthanded, especially when they had
fishing derbies. [We do] things that . . . the Coast Guard would be doing,
. . . except we're not paid. For instance, one morning I was busy . . .
scrambling some eggs in a frying pan. The phone rang, and [this guy] said,
"Lois, can you . . . get your boat over here? We're on search-and-rescue."
I took my frying pan with me and I went, lit the stove on the boat, and then
finished my breakfast out there. We're dedicated to . . . preservation of life
and property.

There's one full women's flotilla . . . back East someplace—Carolinas, I
think. It wasn't that they don't allow any men. There were so many men
and so many women in the same flotilla . . . that they just each have their
own, [which] consists of ten boats. In that particular flotilla, those women
love boating. They're all good boaters. They have the freer time; they're
housewives. I think there's only a couple of them that are working women,
so they can be called up any time for search-and-rescue.

There's one thing that's always bugged me. Somebody's been having a
problem and they call the Coast Guard. They want to be rescued and they're
in a panic situation. So then . . . you go out on a search pattern, which I've
done many times . . . with a close grid . . . not finding the boat, and this
is going on for hours, both Coast Guards' time and mine. [One time] I was
coming into Neah Bay in the fog. I was going to regroup and go back out
and search some more . . . and here was the boat on its way in. Never called
the Coast Guard; never called nobody. Had the same twenty-one, eighty-
two channel we did; could've called and said he'd got going again. All this
time he was busy running. That's happened more times than you can believe.

A man will panic on a boat before a woman will—95 percent [of the time].
One time, off of Umatilla [Reef], there . . . was a fifty-foot pleasure boat
[carrying] a man, a wife, and a twelve-year-old son. I was standing by on
radio watch because the boat was sinking. The Coast Guard was there, and
I was listening to him. The fellow didn't even know how to operate the ra-

dio. They were trying to DF him—you know, take a direction finder—and trying to find his exact location, because he didn't even know where he was. They finally located him, and the seas were a little bit up and down. To get off that vessel, he knocked his wife down and broke her wrist, stepped right over the top of his kid—damn near got him in the drink. The Coast Guard guy said you couldn't believe a man that was just absolutely panicked.

One time, after I got married to Keith . . . his daughter came out with us for a trip. We went in and anchored one night. [We] always leave the—we call it the Mickey Mouse—CB radio on. I was busy making up some gear, and Sandy and I was sitting in the wheelhouse, and all of a sudden it came over there that [this man] he says, "I just hit a rock! Did anybody hear me?" I came back. It sounded like he was right next door, so I says, "Well, it sounds like you're awfully close." I got to talking to him and asked if he was taking on any water. He says he wasn't yet, but he was hung up on this rock. So, immediately then I looked at the tide table, and I said, "Well, looks like you're going to have to get off there sudden like or you're stuck." Because the tide was going out, and that set you right in harder. We had quite a good conversation back and forth. He didn't have twenty-one, eighty-two [channel] to get hold of the Coast Guard, so I called the Coast Guard. Then I found out where he was. [Near La Push] there's a buoy; when they took the Umatilla lightship out, they put two buoys in. Right in between these buoys, there's rocks in here. These flashings on these lights was misleading to him, and he thought he'd passed the second one. [But] he'd passed the first one and turned in here. I was anchored approximately one mile north from him. The Coast Guard asked me to pick up my anchor and go after [him]. I said, "I'm not about to go in there with my boat." So he asks if I had him pinpointed . . . so I gave him exactly latitude and longitude. I knew exactly where he was because I been in there close when it was daylight and charted a lot of area, so if I ever got close in to shore . . . I'd have an idea how far I'd have to stay away from this rock and so forth. Because there's a lot of stuff that's not on the chart. I was talking to this fellow on the Mickey Mouse, then I talked to the Coast Guard on the big phone. And the fellow that I was talking to, he was comical. He says, "By the way, being as you sent out for the Coast Guard, can you send out for a couple of hamburgers, too?" I says, "What do you want on them?" I mean, it was very jolly.

Most people *all* accepted me with no problem. Matter of fact, lots of times after I'd been fishing for about ten years, people'd come up to me and say, "Well, so-and-so told me to come and ask you how to do this." I've done a lot of searching out and finding a person's electrical problems on boats. Part [I learned] in the army; part was my father. I . . . enjoyed helping people out. I'd go over and splice line for them. . . . I'd tell them how to make a long splice. I don't remember anyone resenting me. I never pushed myself on anybody, and mostly they came to me. I used to help out in bay fish office. I'd get free stuff for working, like straightening out gear and logging in catches.

But I did work very closely with Canadians. I had a *lot* of good friends up there that worked together, with gear and stuff like that. We worked back and forth the whole time, and I enjoyed their company. Most of the Americans and their counterparts did, especially with the co-ops, the ones that were in the Canadian co-ops. We were all trying to get the same thing. You help me, and I'll help you; that's the way you did it.

One time [I was] below Tatoosh. It was real foggy. There's a couple of rocks that look like great big fins. In the fog they look mighty peculiar, so I knew exactly where I was. My father had always told me, "Now, if it gets foggy you try to . . . visualize all these things, so if you see them in the fog you recognize them." I saw in between these two rocks . . . this big, beautiful pleasure boat. Then he saw me and saw it was a troller, so he backed up . . . and followed me. I went in between Tatoosh and what they call "the Gut." As you come through the Gut there, there's two wash rocks. The fog had cleared a little bit, and this guy was going to pass me. Well, he'd have went over the top of this great big wash rock, so I motioned for him to get back. They got into Neah Bay and tied up at the fuel dock. These people were parked right ahead of me. These four guys came down the walk there and—I know the guy was going to thank me for it—and the guy stopped short. And he looked at me, and he says, "My God! You're a woman!" They turned around . . . on their heels, got in that boat, and took off. The fellow that was fueling there, he says, "What the devil did you say to them?" I said, "I didn't say nothing to them." It was so funny. I [was] just tickled to death.

[I taught] piloting and navigation [for] the Coast Guard Auxiliary classes [for] eleven years. I would like to see everybody . . . that runs a boat be licensed . . . because there's a lot of people out there that don't know what the rules are. They get out in a boat and they own the water, or they . . . don't know who has the right-of-way. You'd be surprised at the dummies out there. It'd be like you getting in a car. You know what to put the gear in to go, and for reverse, but you wouldn't know the laws about thirty-mile-an-hour limits or whatever. In a boat, you have the same kinds of laws. [When] you come into harbor, it's usually posted two or three knots. Some of those guys come in there flying in these sports boats—anywheres—ten knots. They're responsible for their own wake, but a lot of these people don't know it.

Retirement

In 1974 Lois had to retire from fishing because of a back injury. She tried to persuade her husband, whom she had married in 1973, to take over the operation of the boat.

My husband, the one I'm married to now, he didn't learn to run the boat because he's never been commercial fishing before in his life. I wished he had. You see, the reason I sold my boat was because I was paralyzed for almost eighteen months. I put these Gresen valves in the cockpit for hydraulics to run the gurdies. [In] the rough weather, I would hit back against that Gresen valve so much . . . my lower spine started disintegrating. Then it got so I could hardly move. I have to take spinal-fluid injection shots. If I don't get my shots, then [it] gets so I can't walk very good, and then I have to go to Seattle.

[My husband] did real fine on the boat for handling gear—he did excellent. And he does fish—I showed him how. But he would never run the boat. He'd look at the radar, and he could tell exactly where we were because he . . . knew surveying. So he put the radar and the charts together just beautiful. He'd pinpoint where we were, and stuff like that. But I'd ask him, "Come on. Take the wheel." Dock would be open sometimes when we came in. "Come on. Try it." No, no, no, he wouldn't touch it. He'd . . . change course out to sea, but he would never take the boat in to the dock. I was getting so bad I couldn't walk hardly. I'd get out of the top bunk, and I'd have to lay with my legs over the side for about ten minutes before I could walk on them. That was getting progressively worse. I told him, I says, "Well, Keith, you might have to take this boat in sometime." I says, "What if it got to the point if I fell or something [so] that I couldn't walk?" He said, "Well, I'd just go into shallow water and put the anchor over and call the Coast Guard. They could run the boat." That was it. So then I decided to sell. I sold the following January.

I don't really believe in women's lib. Because a lot of women say—and no can do. I believe in everybody doing what they feel like doing—don't have to have some title on it. [It's not men and women.] It's an individual thing. You never know [who's suited for what work] unless you try. It's like the gal that's that ferryboat skipper. Now that's her cup of tea and she enjoys it. Not all the guys could even take her place. It's just like with me and Keith, a man and a woman—he wouldn't run the boat on inside waters. Now there's an example. You can't distinguish whether it's a male or a female; it's the one who has the desire and is willing to learn how. [If I had had children they would not have held me back], no way. [I would have taken them with me]; that would have been a must. You've got to remember—my mother [and the] tugboat.

After selling her boat, Lois became a fish buyer for a year and tried to start her own fish-buying business.

I bought fish down here for a year, after I sold the boat. That was a fun thing. Man, I was down there on that float with them guys. I was a fish buyer, you betcha. I loved that, but . . . one day I had twenty-six hours in one day. Doesn't sound like very much, but . . . for the gillnet fish, I was getting paid a half cent a pound. On the troll fish, I was getting two cents a pound. So what I had to do was talk the trollers to come here to sell their fish. I paid the best prices. Most of them that knew me came in and sold [to me]. Matter of fact, some guys even put their fish on another boat separate and run one boat in to sell to me, because they knew that I was fair on my prices and I was fair on the scale weight. That's one thing I firmly believe. [I] ask somebody, "Well, how many pounds do you think you got?" If they said they couldn't know, [we'd] check the scales out before we ever started. Because that is the worst thing in the world, is to have somebody cheat somebody. I'd rather it be over than under.

There was some kids I had working down there, didn't know one fish from another. That was that one time we was out there all the way around the clock, and then some. The fish came up in that bucket, and I took a look, and I thought, "Holy cripes! Whose boat did this come off? Didn't come off an Indian boat, did it?" There wasn't an Indian boat down there [and] here I was getting some steelhead. They didn't know what to do with the bucket. God! [If] a fish cop had come down there, he'd have killed me. [So] I buried them in the silvers; I iced those two down in a hurry and got them buried in the silvers. Then when I was over . . . to the place we sent the fish—the cold storage—the boss there said to me, "Did you know that you sent over some steelhead . . . with those silvers?" I says, "There was? Well, I have some young kids there working for me, but I didn't think there was any steelhead sent in." He said, "They didn't end up somewheres else because we cut their tails off immediately." Of course, if they'd got close inspection you could tell, because their skin is heavy—it's like shoe leather. That would be a helluva fine, you betcha. They could have closed me down just quick, right now.

[I did that for] just one season. I was going to build a cold storage, so I invested $6,000 in this cold storage. I was having a lot of static off of the port about getting the building built. [I] went down there and bought anyhow, put a trailer house down there, put in a temporary dock, and had a float with a small barge put in. I was licensed to buy and, of course, I knew all the fish wardens. They was all my friends. Then the port dock really screwed it up.

And there's [a] thing I'm very much against. A port dock should be for the convenience of the people—not there as a business to make money; that's not their principal occupation—making money. They have been knocking these guys as the rates have been coming up and up and up. They've got quite a profit, so they're going to invest in cold storage. They tried to put out the other cold storage. I invested in cold storage here—I wanted to keep right around the fishing—[but] our four fathers and three mothers up there went and put a letter to Olympia not to give us the license. So, there's a lot of skulduggery that goes on in that area. Of course you can't help that; you just have to find out who it is—and take them to another part of the world and drown them, or something.

There's a lot of that goes on . . . under the table. I know. I got some of that loot under the table. I know what it's like. It's mighty good stuff. Of course, it wasn't enough to really bother about—maybe buy a new pair of shoes, or boots, or something—but it was all mine.

[Now, I'm an] accountant. I like to build houses, but I just can't get off work to do that no more. Been building a house [since] 1960—something [like that]—when I got my permit. Come to a screeching halt. I love to build. I like to see things put together and not have a lot of expense behind them. Everything's my hobby with me; I do a lot of sewing. [The Coast Guard Auxiliary work] is all year; it's intermittent.

Although Lois no longer catches or trades in fish, she is still concerned about the future of fisheries, both locally and nationally.

I believe [the Boldt Decision] was a complete farce. I have a friend that's an Indian—that is, in the Indian fisheries—that fished with me years ago. He's over at—they call it "Little Boston" or Port Gamble. He pays income tax because he figures it's his duty. He lives on the reservation now. When he buys anything, he doesn't say, "I'm an Indian." He pays for it. He's one of the very few. He's a hard worker—always has been. He says that some of the things that Boldt—and he and I both knew this, too—that some of the things . . . got carried away, like this 50 percent of fishing. This here, I don't believe in, because the fish hatcheries have propagated fish for world use, fine and dandy.

But it's not only the Boldts. The only reason there's no fish is [the] dumb Russians out there, going clear back [to] when we first started having trouble with the fishing out there. It wasn't the Indians that did this fishing jobby; it was these damn Russian ships and these side trawlers. A bunch of us had been close to this [Russian] vessel, and I took some pictures of it. This [ship] never had a piece of fishing equipment on it. All it had was electronic equipment. [It] could spread its sound out—take a sound and fish here, and fish there, and call their vessels in on it. They'd be there that night if there was fish there. They had fish drying all the way around on that boat—salmon. We watched them pull in the trawl. There was two Canadians that were there, and they was in silvers. Man, them nets was just silver—just glistening. And it wasn't from hake, either, because we saw the stuff when it come up over the top. That's where you lost a lot of your return on your fish.

Our own country's selling us down the drain. I was one that advocated the 200-mile limit. I went to several of the meetings in Oregon and every place else for the 200-mile limit to protect us—not to protect the damn Russians. That's just exactly what they did. Our country turned on us here, because they let the Russians in by permit. They were supposed to have an observer. [But] if anybody has any knowledge of any observer—especially in the past . . . in Alaska . . . in the fish streams—the fish cops . . . were paid off. They let the guys come in there and seine . . . and nobody's going to tell me any different.

I should've probably gotten a little more statistics on this Russian deal and proved [it] to these people, because a lot of these people don't [know]. I had a woman down here at the Carnation Dairy. I mentioned something about the Russians. She says, "Don't talk about the Russians. They're not doing anything wrong." She didn't know. I was . . . telling about how the Russians was taking our damn fish. [She says], "Oh, no they're not." People are not informed properly. A little information is bad; you might as well have none. But . . . I saw all this stuff first hand and knew what was going on. That, I did not like.

I would like to see [fisheries] work for the people. Like [one particular fisheries official], . . . he doesn't know nothing about a fish. He represents for the state of Washington. Now he's in the fisheries. He don't really know which way's up. He's learned some things but he hasn't learned enough. He don't know the difference between a troll fishery and a gillnet fishery. I believe if you're going to have a fishery, you should have somebody that is well represented in the fishery. Say like somebody from the crabbing industry that knows about bottomfishing and crabbing.

They closed off [one area for lingcod]. But what were the lingcod that were in that area? [They] were mostly all males. Now, you figure out. They're preserving males? And a female has 5,000 eggs and up, and she's out there getting caught. It just didn't make sense; somebody didn't know what they was doing. But they get somebody . . . in the fisheries, they go to school without practical experience. I believe it should be a combination of both.

There's . . . a lot of ignorance. It's just like the people that were around where the bridge was before [it was destroyed in the storm]. They'd see the gillnetters in there, and those sports guys made big complaints that those gillnetters were catching their sportfish. And they were on sockeye; [the sportfishermen] didn't know what a sockeye was. Because that's considered a salmon—all those are—and they don't know the difference.

Maybe I shouldn't say that, but you get a lot of schoolteachers and fly-by-nighters . . . out there fishing. [They] come out for the big harvest, or when the bulk of the fish are in. They're not making a living at it. There's something else. I strictly advocate that the fisherman should earn at least 60 percent or better of his living from the sea, or he shouldn't be licensed. Not to go out there and reap the harvest from underneath the nose of the little guy that's trying to make a living at it. That's very important.

[In the past] your gear was relatively inexpensive. You didn't have to go to Timbuktu to get your fish. You didn't have to have every piece of electronic [equipment] on your boat to try to catch fish, and all that stuff. It's just that nowadays you have to be a bit smarter than that fish.

Now confined to the land, Lois still yearns for the sea.

You've heard of the expression "being called to the sea." Well, that's exactly what I had. When I get away from the water, any place, I feel like I'm landlocked. Even after I sold my boat . . . I go down [and] look the boats over. One fellow said to me, "Lois, you look at boats like most men look at women." And that's my biggest calling right there. [I liked] the freedom of fishing; if I didn't like who I was selling fish to, I'd go sell to somebody else. I used to like the cribbage games. I'd play it in the wintertime, when I used to lingcod out there. Used to play a lot of the draggers you'd know'd be in. I'd go play cribbage, and the darndest things I used to play for—apples, pies, biscuits, potatoes. I used to get quite a few groceries that way, because I loved to play crib. I enjoyed the majority of the people. I sure long to go back to sea, I'll tell you.

Katherine (Tink) Mosness

Katherine (Tink) Mosness

Katherine (Tink) Mosness has been around working boats all her life, from the lumber ships inspected by her father to the fishing boats worked by her husband and her daughter. Her early years were spent in Seattle in the 1920s and 1930s.

Early Years

I was born and raised in Seattle. My father was a lumberman. When I was young I spent a great deal of time on the docks and in the woods with him. It was always very enjoyable when we got to go to the various places with him while he would inspect the loading of boats or check lumber mills. Quite often, Mother would pack a lunch that we would have in some scenic place. Then we'd go back to the car to wait for Dad, and Mother would read Paul Bunyan out loud. They were good times.

As Dad was a lumber exporter, he knew the skippers of many of the ships and quite often would bring them home for dinner. We never knew just who he would bring home. It might be a captain from Japan, Australia, Scotland, England, or South America. They were always glad to get off their ship and have a home-cooked meal, and we were always glad to have them. As many of the lumber freighters would have accommodations, they would have their wives along, too. So, Mother had many friends on these vessels, also. We would receive picture postcards and gifts from many places that seemed so exotic to us.

For the past thirty years Tink Mosness has been politically active on behalf of the commercial fishing industry. While her early exposure to boats came through her father, her early knowledge of political activity came from her mother.

My father . . . was not the doer my mother was. She was active in church activities, P.T.A. [and so forth]. Her part of supporting Dad's industry was entertaining the people he brought home . . . and she enjoyed that. Our home was not large, but there was always room for more people . . . and it certainly made for interesting table conversation. She was very active in getting a playground built. She worked through the city planning commission. She never lived to see the playground built. Every time I drive down there now I think, "Well, my mother was the force behind that one." She felt very strongly that involvement would get things done. And I think our family have continued that on . . . through our fishing.

When Tink was twelve her mother died. From then on the family was cared for by a series of housekeepers, and they did not maintain the tradition of entertaining foreign visitors at home.

That was the end of that. Occasionally my sister and I would go down with Dad and have dinner at a Japanese restaurant with somebody, or at a hotel with a captain. But that was the end of . . . the whole entertaining bit. A lot of our friends still had Dad and us over for dinner, but it was never like it was; that was the end of an era.

After her mother's death, Tink became very involved in after-school activities. She was an enthusiastic stamp collector and tennis player, and she particularly enjoyed Girl Scouts.

Seeing we had housekeepers when I was growing up, I probably stayed in school longer . . . than if I had a mother to come home to. I took in all the after-school activities that I could. [I went to Girl Scouts] after school [on] Monday nights. There were thirty-two girls in it. It was limited to thirty-two; we always had . . . a waiting list. We had our patrols of eight girls. It got to be a really big deal if you got to be a patrol leader, because patrol leaders got to meet more often with leaders. You worked so hard to be a patrol leader. What I had done was put my mother image over on my scout leaders. We had scout leaders that were just the most wonderful women in the whole world. They were all friends; they all lived in big houses. They either were up in Washington Park or Madison Park [in Seattle]. It was just a very, very pleasant association. We would do anything in the world for them. We just loved them. That's all there was to it. We had this lovely loyalty. If we needed a little extra help, and we wanted to do something, she'd say, "Oh, I tell you, come Wednesday afternoon and I'll help you with it by yourself." You'd get there on Wednesday afternoon, and she'd have a fire in the fireplace and tea on a tea wagon. Before you did anything, you'd sit and talk for a little bit first. Then you'd get into what you were doing. They spent . . . a little extra time. They weren't as busy as I think we all were by the time we had scouts.

My girl friend and I had the scout troop when our girls were growing up. It seemed like it was so regimented by that time. There were so many meetings that the scout leaders had to go to, and so many district councils, and all of this stuff. Maybe we didn't know they did, but I don't think it was as regimented as that before. I think they gave them a book and told them to go to it. And I think they did. They did the best they could, and we all just loved it.

After high school Tink studied nursing at the University of Washington and then worked for a local doctor.

It was really a hard program, because I was working for a bachelor of science plus my nursing at the time. So, I had everything except the sixteen months of hospital training. I decided I didn't ever want to go into that part of nursing. I liked the science, I liked the academic part of it, but I was never cut out to be a nurse; I'm sure of that. I ended up working in a doctor's office before I graduated. I had never applied for a job. I was . . . [at] my uncle's farm over on Whidbey Island during summer vacation. [I] hadn't even thought about going to work. My cousin called and said, "I want to know if you like such-and-such a doctor." I said, "I don't know. I don't think I know him." He says, "Well, his girl is quitting. Would you like to come back and apply for the job?" I said, "Oh, I guess so." I didn't care, really. I came back on a Saturday, and I stopped in to see him that Monday morning, and [I] started to work that day. I worked there for two years.

After I was through college and working, [my father and I] had an apartment down on First Hill, on Boren Avenue. When I was working in the office downtown, my father and I had an agreement. The agreement was that I would do the cooking and the housekeeping, and on Friday nights he would take me out. So, we didn't go home after work on Friday night. We went to a nice restaurant. This one night we were at the top of the Sorrento. It was a winter evening, and we had a window table. It was very pretty out. We were laughing over something. Some older lady came over, and she looked at me, and she says, "He is old enough to be your father!" And my father stood to his full height, and he says, "Thank you, ma'am. I will tell that to my daughter." We did have a good laugh over that incident. My father and I, we just were friends.

Tink met her husband, Pete, while she was working at the doctor's office. Pete was a patient. She says, laughingly, "As he says, I caught him with his resistance down!" He was not a fisherman at that time. He was a gold miner in Alaska, and Tink went back there with him.

So we went to Alaska on our honeymoon, and we were there six months. We were out . . . seventeen miles from the nearest town. He was prospecting out there. So our honeymoon consisted mainly of being out in the hills. [We used] a Peterborough canoe and native guides; it was very interesting. It was a type of life that I hadn't seen. The gals that were married to miners went to Alaska and lived at the mine. I loved it. We expected to go every spring and stay until fall. Then the war came along and so that changed everything. The price of gold was frozen; it was a nonessential industry. So he started working in a shipyard. He worked as a shipwright for a while. Then he bought a little fish boat, and we moved to Friday Harbor [Washington].

Life in a Fishing Family

Tink and Pete's wartime move to Friday Harbor marked the beginning of their lives together as a fishing family. Two of their three daughters were born during this time.

All during the war [Pete] fished for dogfish, for the vitamin A content of the livers. [It was] . . . a vital industry. So, we had the rations for a crew plus our house rations, which we didn't ever use. We couldn't use [up] our gas rations . . . on the island; there weren't that many miles of roads. Our two best friends up there had farms. One of them was a chicken rancher, so we would trade—as he would say—"dirty eggs for fish." They [also] had fruit trees and vegetables which we traded for fish. [We had] lots and lots of rabbits. We would go out and hunt rabbits for fresh meat, just because it was a change, not because we had to do it. Occasionally we would go out and get pheasants; they had lots of pheasants on the islands. There were just things like that to do. With two small children, you didn't really go very far except to your friends' homes. Quite often our friends from farms would come into town on Saturday night, and we'd all have dinner together. [There would be] a six-pack of beer, and we'd have a party. That was a big night. It was just a very pleasant way of sitting out a war. As far as blackouts or anything like that, there were no military objectives on the island. Occasionally they'd blow the whistles, and that meant that you would turn off your lights and the streetlights would go out. You'd close the curtains, but that was all there was to that. You wouldn't know that there was a war on unless you read the paper. We were out on False Bay the day that the armistice was signed. We had the car radio on. It had [been reported] that New York was going wild and Los Angeles was going wild. We came into town expecting to see everybody in Friday Harbor jumping up and down. There was a dog asleep in the middle of Main Street. We had to drive around it. We probably saw six people in town. [We] said, "Did you hear that the war has ended?" "Yes, we heard it over the radio. That's good, isn't it." "Yes, that's good." But there were very few on the island, percentagewise, that were really affected with even the young fellows going to war. They hadn't gotten that far in drafting that many people.

My husband used to find parachutes, and I'd use the cloth to make things. I made a nightgown one time out of one. It was this pretty white nylon. They were easy to put together to make . . . slips. But the trouble was that they didn't have nylon thread at that time like we do now. To sew on nylon, you have to have nylon thread. We'd sew with cotton thread, and it would pucker. We never were really terribly pleased with what we could make with them, but it was kind of a novelty . . . making something out of a parachute.

After the war Tink and her family moved to Richmond Beach. Her husband Pete became a gillnetter and started building their new home. Later, when all three of their daughters were in school, Tink returned to paid work.

By that time . . . we had three daughters. After our youngest daughter was in school full-time, I went back to work. I worked for about fifteen years in a doctor's office. Well, really two doctors. [When you have] a family, working in a medical office is much easier than nursing, so I never regretted [not becoming a nurse]. I had had business college and I had had nursing. So the combination of nursing and office management was much easier to cope with.

When Pete Mosness was away on long fishing trips, Tink and their daughters continued to maintain the economic and domestic bases of their family's part of the fishing industry.

[In the past] about the first of April . . . [my husband] would leave for Alaska. Then he used to fish down here until the first of December. Sometimes the last day would be Thanksgiving Day. So sometimes he would be in and out during that time and, really, money was tight then. Then he'd have to work in the wintertime. So . . . it was a long season. Of course, we only had one boat then. That meant if he went north, he took the boat up. He ran it up. Sometimes he would fish Southeastern. He'd go on up to Cordova, across to Cook Inlet, and then across to Bristol Bay, and then come all the way back and fish Puget Sound. So they were long seasons and hard, hard work. They don't do that anymore. They have a boat up there. They get in a plane and they fly up to their boat. They have somebody else that fishes with them up there. They just go up for the fishing and come back down here again.

One time back at Boston Fish Expo . . . the home economists had a breakfast and they asked me to be the speaker. I gave this pitch about being a fisherman's wife out in the Puget Sound and Alaska. I ended up with this corny little thing: "When our husbands leave for the fishing grounds, it's like your husband going to work. You kiss them goodbye and wish them well. But our husbands do not come back for sometimes three to six months. All we do is wish them Godspeed and a safe return." I was crying and everybody ended up crying. The gal that was with me, she says, "Oh God, Tink! That was beautiful. That's just how it is." She said that, and everybody came up and thanked us. We went from there into the other room and we watched a movie . . . [about] Winslow Homer. He was sitting . . . on a beach painting. And here was this little tiny boat that was there. The little boy was talking to him. He said, "Why do you paint the boat so small in the painting?" He says, "To give the perspective that the Lord made the sea so big and our boats are so small." And everybody started crying all over again. And that's what it has been all these years.

[When my husband was away, we'd communicate through] letters—just letters. He missed an awful lot [of family life]. He had never seen his girls in the ballet recital. He had never seen his girls march at intermission in the band during a game with their uniforms on. He had never been to Brownie Fly-up. He had never been to the Court of Awards at the Girl Scouts. All of those things he totally missed.

Although Pete Mosness missed a lot of his daughters' school activities, they did all spend time together when he was fishing the San Juan Islands in the summer.

When our girls were growing up we'd rent a cabin on San Juan Island for the summer. We never could buy it, but we rented it for eighteen summers. [It] was on seven acres, out on a point. It was just a shack, but we loved it. We had it for the month of August. Pete would come in in the morning and anchor his boat right there. We had breakfast, and then he'd go to bed for the day. The kids would crawl over the boat and dive off of it, and make rafts, and all the things that kids on the beach do. There were no neighbors, and so each one of them would bring one of their friends from here [Seattle]. We had tents that they would sleep in. Some of them slept in back of station wagons. We had three trees that we had a big plastic tarp around and a sprinkling can; that was a shower. And it was great. All the different boats going by would beep at us. It was a tremendous way for the kids to grow up. They loved their summers.

Tink's description of fishing family life shows the self-reliance and closeness within her own family and between fishing families. They all help one another, and they constantly learn from and teach one another about their business.

[In emergencies] you relied on yourself. [But] nothing [came up] that you couldn't cope with. I think, really, when it comes right down to it, that there are probably closer bonds in the fishing fleet than there are in other industries. You're part of so much of it. It isn't anything [like], "Sign me your paycheck; I'm going shopping." You don't have paychecks. You know where the fish money is scheduled to go before you catch it or ever make it. There are more family decisions, I believe, in our mode of earning a living than there are in others. We're more involved in the actual earning of the living than I think probably you'd be if you worked for a life insurance company. Because if you get fired from that, you could go find another job. Here . . . every cent of your investment goes back in your business. You're involved in it whether you want to be or not. [My husband's] home all winter now . . . and we have constant drop-ins; it's constant people here all the time.

When you marry a man who is in fishing, you eat, sleep, and breathe fishing. It changes the whole family. It's even worse than that. You catch fish all winter. You talk nets; you talk catch statistics; you read reports; you go to meetings. It is constant. People come three and four times a week. You can cut the fish right down the center between all the people talking fish. You live it. The fellows learn from each other. It's a continual education process. Somebody will discuss something in a completely different light than you have thought of it. Now [that] we are having to diversify [there is] even so much more to discuss. Before, you only used your boat for one type of fishing. Now to earn a living you have to diversify. You are getting into things that you don't know as much about. So you are anxious to talk to somebody that has done it, or has talked to somebody else that has done it. It is a continuous educational process. It never quits.

Although Tink has been very active in the politics of commercial fishing, she has not spent much time out on the fishing boats.

I have been out, but . . . with one handicapped daughter I was left at home probably more so than I would have been. She wasn't married until she was twenty-seven. If she had married like . . . the other two girls, around twenty, she would have been on her own probably by eighteen, and I could have gone out. But I didn't. And then I have always had jobs. I also decided that there were other people better on the boat than I was. And I wanted him to have the best help on the boat necessary. [When I did go, I] ate and slept. I didn't work. Down here [in Puget Sound area waters] would be the only time. I haven't been back to Alaska where they fish. Down here they're one-man [gillnet] boats. The fellows very seldom need help down here. Everything is mechanized, and they are given such a short time—twelve hours—to fish at a time. They really don't need anyone down here. So when I go out, I go out for the sheer joy of being out in the lovely evening and taking a nice lunch and being company. That's my contribution to it. Sometimes it's accepted with gratitude, and sometimes it does me more good than it does the fishing part of the family.

The "fishing part of the family" includes Tink's and Pete's daughter Ann. She has followed her father into fishing. It is becoming more common these days for women to be engaged in commercial fishing.

Our other daughter has a fish boat and she's doing her thing. She went fishing with my husband in Bristol Bay for . . . two years. Then she fished Cordova with him; then he went back to Bristol Bay. So, she bought this little boat and was using one of his licenses. To prepare for this she took the different diesel classes and boatmanship classes and things like that. She had been around the boat for a number of years as a boatpuller. [She got involved in that after her first marriage.] She had to find some way to support herself. It looked like something that she'd rather do. So, I stayed home with the granddaughter, and she went fishing. She has remarried and she has a fourteen-month-old little boy now; so I don't know what her plans are going to be for this summer . . . whether she's going to have somebody stay with the baby and our granddaughter. I don't know how she's going to work that. But this is one of the things when the women get involved in something like this. There's always a family that *they're* responsible for if they are married.

[After her divorce] I could take care of her daughter, but now that she's married I ———— . [Her present husband] had a seine boat but he sold it. So, I'm not sure just what he's going to do now.

There are many more [women] in Alaska that fish than there are down here; there are some down here that fish. But in Australia, it was quite a surprise to me to find out how many girls have boats. They do very well. As a matter of fact, the top lobster fisherman in South Australia is a woman with an all-woman crew. It is accepted now. For many years you didn't see very much. You'd see the wives going out with the husbands, mostly for company. There was no reason for staying home if they didn't have a family; so they'd just go along. Lots of wives go out in trolling boats because they are gone for a week at a time. It's not that rugged a life, because everything is mechanized. [They] are not off by themselves. They all keep track of each other, and you have good friends. They're together with other people and all have radiophone contact.

I guess I have always taken it for granted that your backyard is as big as your viewpoint. I think our daughters grew up with that same feeling. As my husband is from Norway, we have had many members of his family here, and I have tried to blend some of the Norwegian traditions along with those that we had in our home, which were basically English. We have had people in our home from all over the world. Many were fishermen from other places; others were government people. It has really given us a broader outlook to our fishing industry that many industries don't have.

Political Activism

Around the time that all the Mosness girls were in school, there began to be problems for Puget Sound area fishermen. It was at this time that Tink became politically involved in commercial fishing. In the thirty years since then, she has been instrumental in the founding and running of a number of significant fishing organizations. These include the women's auxiliary of the Puget Sound Gillnetters Association [PSGA], the National Fishermen and Wives [NFW], the National Federation of Fishermen [NFF], and the Washington Association for Fisheries [WAF]. Tink's first campaign was for the PSGA.

[It all started] . . . in 1954. We had a very bad initiative out here that would have put all our husbands out of work. It would have closed Puget Sound, and we are Puget Sound fishermen basically. It would have closed it except to the sportsmen. We had to do something; it was a frightful thing. Here we were with every cent we had invested in our boats and our nets and our gear. The Puget Sound Gillnetters [Association] called an emergency meeting and asked everybody in the organization to bring their wives. We went up there and they spelled it out, just exactly how bad things were.

There're . . . 600 sportsmen to every . . . [person] in the commercial fishing industry in the state. They not only had us outnumbered with sheer numbers of being able to get to their legislators, but they had the money to back it. All we had was just plain sheer determination that it just wasn't right, that one body of water should and could be shared equally.

There are many types of fish that do not even bite on sportsmen's lines. Those were the kinds that we basically harvest: the sockeye and the chum and the pinks. So they tried to figure out what to do. They said "Well, if we hire this firm, would both the men and the women work with them?" And of course we would work with them. This firm came up with some very nice printed material. Then we didn't quite know how to distribute it. So they suggested that we rent some buses. They'd rent two buses a day. This went on for about ten days. There were both men and women on these buses. We'd get off the bus at this point, and this point, and this point, and this point. We'd take each side of the street, and we'd knock on the doors and ask them if they would take our literature and read it. [We'd tell them] that we were being put out of business, and we felt that if they knew the facts, it would explain things to them, and we'd like to have their vote. They'd thank us for it, and we'd go on to the next house. One day we went to Tacoma; two days we went to Olympia. We went to Everett; we went all over Seattle. We not only got our families doing this, our friends and our neighbors decided that they'd go along with us too. We'd take our box lunch, and we even had a good time doing it. We won so beautifully on that that after the election the fellows asked us to come back to the meeting. And we expected them to thank us for our support and kind of give us a rah-rah-rah for what we had done. But instead, they said that from that time on . . . they had voted that we were a part of their organization. So they sent us in the other room to form an auxiliary. We came up with a group of people [from] the whole length of the Sound that were the nucleus of the women in fisheries.

Everybody felt very strongly that [the fishing industry] had not done a very good job with the PR [public relations] . . . or this initiative wouldn't have gotten such a good start. So we started working in our communities to raise the image of the commercial fishermen. We started having floats in parades. [In] one parade—the Bellingham Blossom Festival . . . we had our big yellow truck. We built the two-by-fours up there and had the net over the truck. We had one daughter in a mermaid suit . . . [complete] with scales sitting on that. [And] we had two of our fisherman gals in sou'westers hanging [and] mending the net on the back. It was fun; we got a lot of good publicity from that.

We would give out recipes also. We were trying to move the last year's pack of canned salmon. During Lent we got together and mimeographed salmon recipes to little three-by-five cards. We would each take maybe a half-dozen stores. Over the canned salmon we would have these free recipes. So people would see a free recipe . . . and it would be right over the canned salmon.

It helped. We [also] . . . put out a very good cookbook.[1] Everybody worked; everybody did something.

The Canadians at that time, too, needed a coast guard. So many of our fishermen went through Canadian waters to go on up to Alaska to fish. So we helped the Canadians form their coast guard. We thought, seeing we share the Fraser River fish with them, that we had everything to gain and nothing to lose from being friends with the Canadians.

The women's auxiliary of the PSGA ran their own newspaper, "The Cork-line," and also provided all the administrative and office support for the men's newsletter, "The Leadline."

We had an office in Everett, and we put out a monthly bulletin. Anybody that could would take their lunch and go up to the office that day. Our paper was "The Corkline." All the news from the different auxiliaries would go there. [It came out] every month, summer and winter, twelve issues a year. We had good communication in those days. The men's newsletter was "The Leadline." They would send in their information—what they wanted out, what their interests were, and what they were doing—and we'd get all the work done. We'd get it all typed up and get it all out—mimeographed, and stapled, and mailed. It was a good community effort and we all gained from it.

Tink Mosness and the other gillnet wives were a model for other groups of women who wished to organize in support of their particular interests in commercial fishing.

After about . . . five years, the halibut wives decided that they wanted to get organized. They called me and asked me if I would come and talk to a group of halibut wives. I said, "Sure." So I came and finally got them started. [I told them] what we had learned and what we felt was the better way to do it. [I] kind of . . . [gave] a sales pitch to get them going. [I told them] how it would help them.

The gillnet and halibut wives then went on to help the troller wives to organize and, later on, the seine wives. The women of the gillnet, halibut, and troller organizations founded the National Fishermen and Wives [NFW] in 1963.

1. Puget Sound Gillnetters Association auxiliary, *Seafood Treasures and Selected Restaurant Recipes* (Lenexa, Kansas: Cookbook Publishers, Inc., 1978). Available from PSGA, Fishermen's Terminal, Building C-3, Room 103, Seattle, Washington 98119.

Then my friend that was the president of the troller wives called me one day. We had talked about having an organization for the men, like the Grange for the farmers. She says, "I think the time is right." And I said, "Fine." So she said, "If you will come to our annual meeting tonight . . . we'll see if we can put something together." So, at that meeting she had me give a pitch, and she had [the president of halibut wives] give a pitch, and she gave a pitch. We asked for a show of hands if the men at the meeting thought that it would be a good idea. Of course the very last one of them, if somebody [else] wants to do something, they'll all clap for them and have them do it as long as they [themselves] don't have to do it. So, we came up with this [plan]. We'd start a bank account called the "Fisherman's Fund" in the Ballard Bank, and everybody that thought it was a good idea would either mail in . . . or take in a dollar and put it in the fund. When we got to $500, we would know then that enough people felt that we should have a national organization. We thought maybe it would take six months. Well, within three weeks we had our $500. The three organizations formed the nucleus of the National Fishermen and Wives. Little by little it grew with other organizations.

In 1969 the National Federation of Fishermen [NFF] was formed through a merger with the Congress of American Fishermen.

Then another organization started. It was the Congress of American Fishermen. Another fellow that owned another fishermen's newspaper went to the seiners. He said that he would give them all the [newspaper] space they wanted if they [would] start an organization. So, they started another organization instead of making one organization [with NFW]. It was just a headache, because the other organization had a newspaper that would do nothing but blow up what they had done. It was kind of discouraging. People didn't know which organization to join. They'd get all this publicity from [the other] one. [The other group] started out with a great big bang. But they wouldn't have women in theirs [whereas] we had the men's organizations and the women's organizations. We had the big tuna boats from San Diego; we had the little gillnetters from Seattle; we had the auxiliaries—we had everybody—and it was the National Fishermen and Wives. They decided to merge the two organizations. We changed it to the National Federation of Fishermen. Our main purpose, of course, was to have the East Coast and the West Coast in one organization, and to have the Gulf Coast, and eventually to have a Washington, D.C., lobbyist.

The success of Tink Mosness and other women in the NFF resulted from their use of multiple strategies for tackling one problem. For example, the Pacific Coast fishing fleets were being harassed by foreign fleets. The women in the NFF went to Washington, D.C., to lobby for a 200-mile limit, and they also created their own, more immediately practical solution.

[During the time] they were having all these troubles with the foreign fleets . . . the fellows would call in to a Coast Guard that they were being harassed. The Russians would immediately pick it up on their radio. They were monitoring everything. By the time the Coast Guard would get there, there'd be nothing. So, they came to the federation and asked if there were something we could do. We came up with a code system. We had a list of boat parts. All the boat parts were in code of a country, size of a vessel, where they were fishing. The number of the boat part was in a grid on a map. We did this up so that every Coast Guard station all the way down the coast to northern California, and all the fishermen's wives along the whole strip, all had these same books. If one of the fellows on a boat would call his wife, as if he were ordering boat parts, she would take this information down as he was reading it from this book. Then she, in turn, would pick up the telephone and call the Coast Guard. They, in turn, would have the same book and would then know without the Russians, or the Japanese, or the Koreans—or whatever—knowing. They would get help out there immediately. We had a communication system. We were . . . solving problems that we were able to solve.

Although the foreign fleets didn't affect our little gillnetters, they did affect the trollers. The trollers' wives asked us for help. Eight of us went back to Washington, D.C., with a salmon that had a Russian net on it. [The net] had feathers tucked in. We have no high-sea net fishing for salmon; we have no nets with feathers. So we had this salmon in formaldehyde [and] we met before the Commerce Committee. [People in D.C.] said we'd make a much better impression if we came back there as if we were representing a business . . . which fishermen are. They said, "If you are testifying, look like businessmen's wives, not fishermen's wives." So we all went back there in our high heels and our hats and gloves with this pickled fish! We went back there just before the delegation left for Tokyo to meet with the international meeting on this. That story came across. We didn't get the 200-mile [limit, but] they did pass the 12-mile. Congressmen Pelly sent . . . a cable to the plane [as we were coming home]. They brought it back to us. Then they had to put over the PA [public-address system] what we were all celebrating for. That was the first time that our fishermen at least had protection more than 3 miles out.

It is worth noting that much of the significant lobbying in Washington, D.C., has been done by the women's auxiliaries, rather than by the men.

Women can get together and go together as a group. All organizations now have big paid directors. If they go back there . . . they'll take an attorney or they'll take another fellow with them. But they go back as their own. They go back there to represent their organization. They're representing *their* viewpoint on that one thing. The seiners will send somebody to Washington; we'll send somebody to Washington; the trollers will send somebody to Washington; Grays Harbor Gillnetters will send somebody to Washington. They all testify as individual organizations. Women don't have to do that. We can go back there. We all have a viewpoint on how this is affecting us, but we're all together. We're fighting for a way of life. We're fighting for a livelihood. [The men] don't work together. I think that's where they lose ground [and] where we gain. We have a better communication system. We all know each other and there's a bond between us. Our kids all knew each other and we had the same interests. When we have gone back there, many times, we have gone back as a group representing a viewpoint. The fact that we are from different places, different types of fishing, is a story in itself. We can get people to listen. [The men appreciate our accomplishments]. I'm sure they do; I know they do.

The fishing organizations in the Puget Sound area were unable to get the NFF to support them in a stand against the Boldt Decision. This came as a blow to Tink Mosness and her coworkers who had fought so hard for a national organization that would represent the interests of all the fishing communities.

So many of our organizations are not in . . . [NFF] now, and it is over one thing. We feel very, very badly about it. We have a very, very bad situation out here, and that's the Boldt Decision. We worked so hard to have [NFF] represent everybody so that we would have clout. But we don't *have* clout because we do not represent everybody. I feel very strongly, if we had had this organization backing us during this Boldt Decision, we could have gotten this cleared up. But we were fragmented. The other organizations did not back us because we were not in the organization [NFF], and we couldn't back an organization that would not represent our problem. Twenty-four of us [women] went back [to D.C. as representatives] of different fisheries organizations [to lobby against the Boldt Decision]. We got the group together . . . from the coast of Washington, the coast of Oregon, and northern California. We wanted to meet with the presidential task force before they made their recommendation to the president. We wanted to show that this is not a local problem, that this problem is a blight affecting the fishermen all down the coast. That this is a federal problem, and only the federal government can solve it. We talked to the Department of Interior; we talked to the Department of Justice; we talked to the Department of Commerce. We met with all our senators and legislators. Nobody had *ever* gone back there and presented it as a whole coastal problem before. But when it comes right down to it, we still lost.

Because of the problems with the NFF, the Washington Association for Fisheries was formed.

We now have the Washington Association for Fisheries. These meetings are held . . . monthly . . . at one o'clock in the afternoon down at Fishermen's Terminal [in Seattle]. This is husbands and wives or anybody. We don't divide up men and women; we seem to have outgrown that. We have gone back to a statewide organization that is working on state problems. We do have a good strong organization with that now. We have trollers, we have gillnetters, we have seiners, and we have sportfishermen all in the same organization. We're working against the Department of Fisheries doing some of the things that are affecting the different segments. We're working for water quality; we're working for the Public Health Hospital; we're working on constructive things that affect the fishermen of Washington State. But we really should be working on those on a national scale.

One of the major concerns of the Washington Association for Fisheries is that salmon resources in the Puget Sound area should be managed in a way that will ensure a plentiful supply, so that the fishermen can make an adequate living.

The Washington Association [for Fisheries], when the Fish Expo was here [in 1979], had a forced-out-of-business sale. It was strictly for the press. We had these forced-out-of-business signs. We put them on every boat in the fleet. We got the press down to show that the whole fleet was for sale. These people come [to Fish Expo] from all over the world to sell their wares. One whole segment of fishermen can't buy their wares. There weren't any people fishing Puget Sound, and very few fishing the coast, that were buying what they had in the past. We wanted to show this.

Although Tink Mosness is no longer a member of the PSGA auxiliary or the NFF, she is still very actively working for good fisheries management and for the health and welfare of the fishing community.

I have a couple of other interests. When I was secretary of the National Federation of Fishermen, they started to close the Public Health Hospital. So I was sent as health and welfare [representative] of the national federation to this Public Health Care Coalition representing . . . fishermen as the beneficiary group. Now that was eight years ago. I still not only have that job on the coalition, which is one meeting every month . . . but . . . now we have a Patients' Advisory Committee of the Public Health Hospital. I am an appointee by the director of the hospital on that. We're really accomplishing something that is [of] benefit to all the fishermen.

[I'm also involved with] the Interstate Congress for Equal Rights and Responsibilities. That's a national organization in twenty-two states. It's to get
back to the theory of the Constitution that we're all equal. We are not equal
in this state now. We are second-class citizens to the Indians as far as our
fishing and our property rights go.

*Tink Mosness has become involved in arranging and accompanying tours of
fishermen to Australia. She envies the influence that Australian commercial fishermen have in the regulation and management of their industry, and she is critical
of the policies of the Department of Fisheries in the Puget Sound area.*

I just took a group of seventeen fishermen down to Australia. They had
brought sixty-six up here. We've been doing this back and forth for a number
of years now, exchanging [with] and learning from each other. [When] we
came home from Australia this last time, we were ready to roll up our sleeves
and take on our government. There's nothing that the government won't do
for the fishermen down there. They don't call them "commercial fishermen." They are called "professional fishermen." They are on advisory com-
mittees. Every advisory committee over in Australia has commercial fishermen from every single area. [They are] into the management, into the
regulations, into everything. Here [government officials] write the regulations and mail them to you. That's it! We came back really feeling that the
fishermen had the respect of the government down there, which we really
don't have. We're kind of outcasts, and we don't like it.

If we can get our state Department of Fisheries out of politics, there is no
reason we can't get back to a good sound fishery again. Fisheries is so involved in politics in this state. The fisheries department has come up with a
coho salmon that is supposed to be for the sportsmen. They keep it until a
late release here in the Sound. Pete has caught some of these. We have pictures of when you split them open; the belly is filled with baby salmon. There
is no feed for [these coho]. They don't go out to sea like the other salmon do
and range [for food] clear out in the ocean. They're staying here, supposedly
for . . . the local fishermen to catch. They're eating the fry. So it's a point
of no return. We're never going to get [problems] like that [under control]
until we get the fisheries department *out* of politics. A number of things . . .
could be done. We can raise certain types of salmon so easily. But if [these
things] aren't [done], I'm not sure that there is going to be much of a salmon
future, either for trolling or here in the Sound. The main [place] is going to
be Alaska. It hasn't gotten that bad up there yet. But [in Washington] we're
involved in politics.

Now the Indians are fighting with the sportsmen; the Indians are fighting
among themselves. The Indians are fighting with the commercial; the sportsmen are fighting with the commercial. There just is a point of no return. We

could have enough salmon to keep everybody happy if we could get [fishing] out of politics and get back to managing. The outlook [for fishing] isn't that good that it entices people to start with a little tiny boat to go out. It used to be cheap to get into fishing. It *isn't* anymore; it's expensive. It's something which you would mortgage your soul to get into. [Right now] if you are going to earn a living in fishing, you might as well get into it and learn to fight. It's not a nice little thing like farming, where everybody stays in their own little fences. You're overlapping with other fisheries; you're overlapping with other water users.

A lot of the problems are from a lack of information. [Others] don't realize that we're not the bad guys, that we can all live together. Even the loggers, we fought them very [hard] on this clear-cutting. They had ruined so many of our rivers and so many of our streams. But by working with them, now they clear-cut two hundred feet from the navigable water, and they quit. A lot of our silt problems have been resolved. We have worked *with* the different chemical companies and pulp and paper companies, trying to get the enforcement of cleaning up their pollution and their soot and things that they're putting in. But we have [only] one-half as many bodies . . . working as we could have. There are still so many things to do. But it's been a good industry. It's been a good life-style. We have enjoyed it, and it has been good to us!

Evie Hansen

Evie Hansen

Evie Hansen was born and raised in Aberdeen, Washington. Her father was a fisherman and her mother ran the farm on which they lived. As an adult Evie has become intimately involved in commercial fishing, both as a fisherman and as a political activist. As a child, however, her life was centered around the farm.

Childhood in a Fishing and Farming Family

[My mom and dad] were married during the Second World War. When
my dad got out of the service . . . he decided that he wanted to be a fisherman. So they moved to Westport, Washington, and they bought a boat,
and he went to Alaska and went fishing. That was the only year he went to
Alaska. It was a terrible season. Then he came down and just trolled off . . .
the Pacific Coast. From Westport they moved up the Wishkah Valley [near
Aberdeen] and, along with fishing, he farmed. It was just a relatively small
chicken farm with just a few thousand chickens. We grew all our beef. [We
had] . . . our huge garden. We canned foods and preserved as much as we
could.

I can only remember bits and pieces of what fishing was like [while growing up]. I know there was always an excitement coming about—oh, March—
maybe before that. It was the anticipation of the season, and we never knew
what our season would be like. Maybe it would be a good season and maybe
it would be a real bust. But there was always the hope that it would be a good
season. And there was always the discussion of "Maybe this year we'll get to
Disneyland," or "Maybe this year we'll get to build our new house, or buy
another car." The anticipation of the season, I think, is what I enjoyed the
most. I remember it the most as a child, hoping it would be a good one. And
I still think, to this day, that's what I like about fishing—the *anticipation* of
it.

[My father was away fishing] from the first of April until the first of December, and possibly longer. He would be gone then for weeks, months at a
time; he would call home and check up on everybody. [My mom gave birth
to three children.] Dad was home for Steve's and then for my birth. He was
in Alaska at the time when Doug was born, and [so] she went down to her
mom and dad's to have my brother.

[Life was difficult for Mom.] I remember my mother getting sick. I think
she had hepatitis . . . and it was terrible to see her just laying there in bed.
She had to hire a person to come in and help, because there were three of
us kids and the chicken farm, and it was in the summertime when she had
to get the vegetables in the freezer and everything canned. I think she probably had just pure exhaustion. Dad was gone. [But then] the other thing was,
when he was there . . . Dad didn't do anything in the house anyway. He
was from another type of life-style where the men didn't do anything in the
house. They brought home the bacon, and then from there it was for the
women to do everything about the bacon. Everybody [had to help out on the
farm]. Lord knows, I've debeaked millions of chickens in my lifetime, and
plucked a million of them, and cleaned eggs, and shoveled you-know-what
out the door. [My mom] didn't have any comforts of home. She had a wringer washing machine for years and years. I remember her telling us never to

touch the wringer when it was going because our arm would end up in it. And she didn't have a drier. I can remember the clothes hanging up on those wooden racks.

She said it was really hard, and it's one of those decisions that you make in life. She knew she wouldn't get a divorce because of it. So then she had to live within that decision, since she looked at it as a lifetime decision. And she really found a lot of comfort . . . from the Lord and from prayer, because there were times when she was at her wits' end and she just needed somebody, heavenly or otherwise, to say, "Hey, it's okay—you can make it." And she told us kids over and over again that it really came from the Lord, that she really received—still does receive—a lot of comfort from Him. So it wasn't something that she really felt like she could do herself—pull up on her bootstraps and make it—because she did suffer physically . . . from the hard life.

[My mom's friends] were neighborhood gals, and they were all farmers. So they could share that kind of life-style together. She never did share much of the fishing at all with anyone else. There was a couple that I remember now. They were a fishing family and we did share a lot of things. But there was no organization or anything like that where she could share the troubled times with anybody else.

Evie attributes her belief in the value of education and her religion to her mother's influence, and her preference for independent living and participation in the commercial fishing industry to her father's influence.

My mother and I can share really deeply where we're coming from, and how it's been. Her biggest prayer for us when we were growing up was that we would not marry somebody who wasn't a Christian. She felt . . . she married against a real principle that the Lord has given us [which is] that we "be not unequally built together." [This] doesn't mean that it's going to be utopia when you're married to somebody who is a Christian. It means that you've got a real common basis for a life. She gave me a background for religion, which is what I've always followed.

[I am close to] my mother, but I wouldn't say that I was really terribly close with my father. [Dad] is really independent and he's self-made. It's really hard for him to share much of anything besides the weather and the fishing season. [Although he isn't a religious person,] he expresses himself at certain times. Like, "I was very thankful that the Lord brought me home," after a particular storm. He went out and he was worried about it, out in the ocean, and he tried to call the Coast Guard for help, and his radio didn't reach out. He had to do it alone, and he realized that the Lord was with him then.

[The values I've gotten from my mother were an appreciation of] education and of the finer things in life. I enjoy good music and different kinds of music; I enjoy cultures. [From my father, I got the values of] being independent. I *abhor* being stuck in an office day-in and day-out. I like change and I like challenges. When you have an eight-to-five job, there's really no risk involved at all. You know that your benefits are such, and your salary is going to be such, and there's really no hope of ever gaining anymore than that. But with [fishing], then you take the risk of not gaining anything at all; when you have no set income, you can lose too. [Fishing is like a gamble], that's all it is. I think somebody said, "I hope your life is full of windstorms." Because it's when it's windy that you learn a lot pertaining to the wind on the water; when you're being challenged to the utmost. I like the gambles.

[My father still fishes], although he's really slowed down and kind of cruises. [He knows the pattern so well that] he knows when the fish will be out there in their fullest forms. So he goes out and goes fishing when it's fairly good— and then he ties up the boat. He's put in his years; he's seen enough of the fishing that he's ready to just kind of sit and mind the boat. [My mom and dad] are on a trip now to California. They're just kind of laid back and enjoying life.

My middle brother, Doug, and I are . . . extremely close; he's a fisherman. My oldest brother, Steven, chose the eight-to-five routine, and we're just opposites. Doug fished with my father, so he got an early taste of the independence. Steve never did care about fishing. Although he was asked to fish by my father, he just didn't like it. He didn't like the uncertainty of fishing. Doug has his sailboat and goes fishing commercially in Alaska. He also fishes out of Westport and Puget Sound.

Before the Boldt Decision

Evie Hansen went to Gray's Harbor Junior College and Seattle Pacific University. She earned a degree in special education and then taught for eight years in the Seattle public schools. During that time she met her future husband, Randy.

I was in college at Seattle Pacific and a roommate of mine had met Randy's cousin at church and he said, "Hey, you'd better come meet this guy. This Randy critter is a fisherman; you guys will have lots to talk about." So it took quite a while, but finally I decided I'd go and meet this guy. We were married two and a half years later [in 1970]. We had some rough spots along the way. I remember [that one of our disagreements occurred one time] when he came back from Alaska. He'd been independent for three or four months and he didn't want to come back into a dependent type of relationship. He didn't want to get back into the dating thing at all. I think that is something a fisherman's wife goes through now, even. Where, for three or four months, they're off doing their own thing and are independent, [and] we're at home with the kids making decisions. And then, all of a sudden, you have to start functioning again as a unit, and there's all the adjustment.

[What attracted me to Randy was the fact that] he was always a happy kid and had a lot to offer. He is also a Christian, which was a priority. [The fact that he was a fisherman attracted me also.] My family likes the idea of Randy's being a fisherman, but they also like his character. His father's father fished in Norway, and they came over here in the late 1800s, I believe, or maybe earlier. They built boats and fished. His father and grandfather build boats and fish now, too. They've done salmon fishing in Alaska for many years. Randy started fishing when he was six or seven years old. He knew nothing else and, as I said, his dad was either building boats or fishing. He was on the boat a lot as a kid, as a paid crew member.

[Our marriage is] really a good friendship; yet it's something that we work at, too. It's not something that we just take for granted. We evaluate it, kind of look at it, and say, "What direction are we going now?" [On the whole, Randy and I] can have our differences about a certain opinion, and yet I respect him and he respects me. Some husbands and wives don't have that sort of respect for each other. The man is still the dominant table-pounder; if the woman has a different opinion, then she's just beat down to nothing.

When Evie married Randy she began to fish as well as teach. They both taught in the winter—Randy in fisheries at the University of Washington—and went fishing in the summer.

[When we were first married we did not own a boat.] His mom and dad owned a boat, and we would lease it from them. We traded boats with another guy so that he could . . . [try out a] troller . . . for the summer [of 1972]. We had decided that we wanted to find out what trolling was like, since my dad was a troller. We decided to go out to La Push, which is the only safe harbor when you leave Neah Bay . . . out on the point of Washington. You've got Neah Bay, and La Push, and Westport to duck into if there are any kinds of storms at all. And lots of times you can't get into La Push; it's too tricky to get into. So I went there to go trolling with Randy, and I didn't like trolling at all. It was fourteen or eighteen hours a day, and we're out on the open ocean, and you didn't see land for hours at a time. I hated it. I said, "Well, there's got to be something I can do." And, sure enough, they needed . . . a summer schoolteacher on the reservation [at La Push].

I was given a huge amount of federal money to make a successful summer program on the reservation. So, I centered it around basic skills and did that for two years. [I enjoyed the job,] except that . . . I was recognizing where it was a buy-off. Unless the kids—unless any of us—really want to learn something, you can't shove it down their throats; you just can't. The tribe thought that maybe by having it on the reservation, the kids wouldn't have any identity problems. There wouldn't be any prejudice or anything like that. [But, the kids] could look out, and they could see that when they grew up to a certain age they could fish whenever they wanted to. What value is education then? It's of no value. At this time, in 1972, before the Boldt Decision, the Quileutes were selling steelhead to the state of Oregon and they were making hundreds of thousands of dollars. Those kids had a taste of gold.

After one season of trolling, Randy decided he preferred net fishing and began gillnetting in the San Juan Islands. Evie taught a second summer at La Push while Randy gillnetted. The following summer, she joined him on Lopez Island rather than return to La Push. She began fishing with her husband more often.

[I fish with Randy because] I like being on the water. I sort of take the easy way out, though. I've had a trailer on Lopez [Island], and if it's been really, really rough, then I won't go out fishing with him [because] I don't like the super rough water. It doesn't bother him at all, but I don't like to bounce around that much. We don't have any children. We can lock the house up and go and don't have to worry about getting home because the kids need school clothes, and school starts, and that sort of thing. I do a lot of navigating and . . . pick fish when there's lots of fish. Usually one person can pick the fish most of the time, but when you're in real heavy fishing, it's nice to have people with whom you can get the fish picked and bring it out again; it's a lot faster. [I] . . . cook and keep the boat up and clean it up, although Randy probably keeps it cleaner than I do, so I can't say that I do my share of that at all. [I provide] . . . companionship. I think that's one of the things that happens to a lot of fishermen—the loneliness. And then they rely on their fishing buddies over their families. I think that times apart are really healthy, but you can't have a lot of time apart and have a real good relationship, either. So, we have a little bit of both. You see them too much, too, when you're on the boat. It's just a little tin cubicle, and you're rubbing shoulders all the time. That can be too much.

Because fishermen are away for extended periods of time, and families are separated, there are close friendships within the fishing community.

[Our circle of friends includes fishermen, fishermen's wives, people from church, and a lot of family friends.] It's not very hard [to meet other fishermen]. When we were in Hawaii—you wouldn't believe it, but on the plane going over there were fishermen that we didn't quite know very well. Somebody would say, "I think I've seen you on the dock, haven't I?" So, you just sort of meet people [and when you see each other] you have so much to talk about. It's . . . like when you see two mothers in the grocery store, and they each have two-year-olds, and pretty soon they're talking with each other. They've never seen each other before, but they have this common bond. Not that you're wearing a sign that says "I'm a fishermen," but if you've seen someone, or they look a little bit familiar, there's always a connection somewhere. It opens up the conversation and starts friendships that way.

The Boldt Decision

In 1974 both the Boldt Decision and legislation to improve the socioeconomic position of ethnic minorities affected Evie's teaching and her and Randy's fishing.

When the Boldt Decision came along, all of a sudden I was seeing this encroachment of federal regulations. It was in '74 when the state legislature put a moratorium on . . . licenses. We knew we had to get a boat then, because there weren't going to be any more licenses issued. So we signed a contract for our boat and then the Boldt Decision came the same year. Here we had this contract, and the Boldt Decision said that we couldn't fish anyway. And what we were going through in teaching were mandates from the federal government, because special education is federally funded. We had all these special requirements to fulfill—paperwork like you wouldn't believe—coming from the federal government. And then Randy and I were fighting the federal government through fishing, and trying to figure out what in the world all this judge business was all about, and how . . . [a judge] could make one decision that would affect our lives so dramatically. Then there was also the hiring and firing of teachers because of quotas. Quotas were being set, and [in] a lot of cases [were set] by race. One year I got a little white envelope from the principal. It was marked "Confidential." I opened it up and here was this form. It said, "You can fill out what you think you want to be, what you feel you are, and mark X here." There were seven categories. Gypsy was one of them, Caucasian was another, and Native American, and so on. So, I decided I'd just bring myself out and be an Indian. I thought, "If it's come to that, I think I'd better hang up my hat." So that's when I quit [teaching] after eight years.

As well as fighting the Boldt Decision itself, the commercial fishermen found themselves fighting media images of the conflict over the Boldt Decision.

[Something] the media [has] tried to do over and over again is make it into a racial issue, where it was like the black and white fight—the thing of the sixties. They were trying to put the Indians into that category also. Here there was going to be this major fish war, and it was going to be [between] Indian and white fishermen. And race had nothing to do with it at all. We still want to have solved the issue of: "Can citizens of the United States have treaties with the Indian states?" That's the major underlying issue. What's happened is, it's gone from fishing into land, and to water, and other issues that can't be ignored. [Like] when the whole state of Maine is being taken over by the Indians——.

[As far as the media goes] when things were at a real smoking point, we decided that it was time to do a little PR [public relations]. We took a beautiful sockeye down to [a local TV newscaster]. We told her that we wanted her to have this fish, that this was something that stood for what it was all about. Her comment came out so quickly that it told me where she was coming from when she said, "I hope this fish wasn't caught illegally." I thought Randy was going to hit her, and maybe he should have. [I saw that] she does have a lot of power as a media person. I don't respect people who don't have an honest view. I can honestly say that I can see both sides of the picture. I've lived on a reservation and know where the majority of Indians are coming from. So the power of the media is something we're all struggling with. I think we're all finding out that there is . . . freedom of speech [but] then there's also the imposition on my rights as a person and as an individual. We're still [finding] the residue of the liberal type of approach, where the poor Indian is the underdog, and you've got to promote him to some higher levels by knocking this other fisherman off the water. That was basically the way the media approached it. We had [one politician who backed us]—Jack Cunningham, who was defeated soundly after he supported us. I think most of the politicians have stayed away from us [because] . . . the media tried . . . to make it into a racial issue.

The Boldt Decision had strong implications for the way that Evie and Randy could fish.

[Before the Boldt Decision] we had a lot more fishing time and a lot less regulations as far as gear restrictions, area restrictions. We could fish way out in the Puget Sound, or way inside Puget Sound. It could be our own decision where we wanted to fish. That was part of the gamble on making the right decision or making the wrong decision. But now there [were] . . . maybe seven days of sockeye fishing this year [1980] compared with forty, fifty days that we [had] before; and in those seven days, the area was restricted. We had a real small area to fish in, too. So, we've just been regulated down to no fishing at all; essentially that's what it is, because you just can't make any money at all in that short time.

As well as affecting the Hansens' fishing procedures, the events surrounding the Boldt Decision affected and changed Evie personally.

I don't know that I even voted all that much before the Boldt Decision, but I think all of us have become cheap little dime-store attorneys. [I] look at things a little bit differently than I had done in the past. Things like "injunction notices" and "writs," words that I had never even heard about before, suddenly became part of our lives.

[Since the Boldt Decision] I have become more outgoing. I think I have something to give. I can ask a question and not be embarrassed because I asked it. I can remember going through high school and college thinking, "I can't ask that. That's a really stupid question. I'll be laughed at." Now, I don't really care if I'm laughed at. I want to know the answer. I think I had a small-town personality, where [you] just sort of sit back and let other people do it. And now I'm willing to do it if I can. I've recognized where my role is in that. One of the things I could see in the auxiliary [of the Puget Sound Gillnetters Association] was that I could draw people together.

I tried to do something constructive with my anger, like becoming a lot more active politically than I'd ever been. Whether it's constructive or not I don't know, but I try to do something with it, like writing or calling our congressmen or our senators, someone who can do something about it. I guess that's where I see the checks and balances, the kind of government of the people. When the federal marshals were coming . . . to our doors night and day, pounding on them to hand us injunction notices, I was extremely angry. They tried to run us off the water, and now they were at our homes. I was calling newspapers and TV stations. [The] TV [people] were coming out and they were showing the federal marshals getting a 92-year-old fisherman up out of bed.

One of Evie's responses to the Boldt Decision was to join the women's auxiliary of the Puget Sound Gillnetters Association [PSGA].

[I joined the Puget Sound Gillnetters Association auxiliary] after the Boldt Decision in '74. [I joined because I knew that] unless we banded together we'd be sunk, and I felt like there was safety in numbers. So the more people that we could get [involved in] a certain project or aware of what was going on [the better]. The old divide and conquer trick was being used systematically. They were playing purse seiners against trollers, for instance. I wanted to change that.

The auxiliary went to Washington, D.C., in '77. There were twenty-five gals and they were from Washington, Oregon, and [northern] California. We picked people up as we went along the way. The auxiliary was the main organizer. Actually Tink Mosness was the one who organized most of it and put it all together. She was [also] a travel agent, so she could get reservations, such as they were, at the Burlington Hotel. At that particular time [when we went], we were really hurting; we'd just been put down to zero as far as fishing time. The other fishing groups [like] the trollers still weren't convinced that they were going to have anything to do with the Boldt De-

cision. [Some of these women felt] they were being used for the Boldt Deci-
sion issue, because wherever we'd go and whoever we'd meet with, it was the
Boldt Decision [that we discussed]. But they realized [through this] that they
needed to have some unity, and [so afterwards] we formed the Pacific Coast
Coalition of Fishermen's Wives. They have since been impacted by the Boldt
Decision. So now they're really into it.

[As for the trip], oh boy! What an education! I think that time was my
lowest point [and I had a lot] of cynicism. It didn't matter who I looked at.
I saw them as federal employees who were more concerned about their benefits and their retirement than [about] what they were doing. The highlight
was when we crashed through the Department of Justice. [Because] they were
so concerned with their security, they had police standing at every doorstep;
you were searched as you went in. I have to say that we never lied, but that
we just said, "There's some people meeting in Room 212, and we need to
get up there. We're late. Can you tell us where the meeting is?"—not saying
that we weren't invited, but that we needed to get there.

We walked into that meeting. There were about twenty-five bureaucrats
who were deciding our fate, deciding when we'd fish, and when we couldn't
fish, and how they would issue injunction notices, and so on. [There was]
utter silence when we walked in. They couldn't believe that we would come
and stand there and tell them that . . . we were people, that we actually
had families, and boat payments, and actually had needs. All they were concerned about was: "Did the poor Indian get screwed back in 1854?" I think
that was the highlight of [the trip]. [During the trip] the attorney general told
us, "By your fighting, whichever way you choose to fight, you have gained
years on your livelihood." The Department of Justice didn't have any idea
that we would resist as much as we did. By resisting [we gained some time]
. . . to establish ourselves in other fisheries and get out [of gillnetting].

*The activities of the PSGA have waxed and waned in response to particular
political events.*

[Looking] back over the years, when things have needed . . . to be taken
care of politically, the group has been active. We've had four or five years of
intensity, and now the auxiliary is kind of backing off. [This doesn't bother
me. In fact,] I think that's good in a lot of ways. It's good to get a refresher
and get off into something else. For instance, I've enjoyed doing the thing
with the cookbook, which was really fun for me because I've gotten into marketing and finding out how people compute percentages and figure out their
profit margins. Right now we [PSGA] have elections coming up in January,
and it's even hard to get people who want to run. There have been years in
the auxiliary in the past where they haven't done anything. But we'll always

have [Fish] Expo every year, [and] we'll always have the flea market, which comes up in April.

The women in the PSGA auxiliary have provided unique and vital political support for their livelihood, which most of the husbands appreciate; a few husbands, however, do not.

Some women . . . can't be part of the auxiliary because their husbands won't let them, and we understand that. Some men don't particularly care for women in any position; they'll talk about the woman governor [*former Washington Governor Dixy Lee Ray*–ED.] as they will talk about the auxiliary. We've had some run-ins where they haven't liked what we've done, and we've come on without them. That's their problem, not ours. We haven't liked what they've done, either, in a lot of cases. Yet our goals are the same; there's really no conflict as far as our goals are concerned.

[Randy is not as political] as I am. He gets angry and . . . that can block communication. [We're a pretty good team,] where I can be pretty aware politically of what's going on and share information with him; he digests it, and we both discuss what we can do about it. He'll write letters and stuff, but he won't be the kind that'll be active in public. I think the men just don't want to spend time on [politics]. They're so tuned to hanging nets. Women recognize the needs of an organization more than men do. If there are different types of fishermen, we all have common needs. One thing that I got from the coalition [Pacific Coast Fishermen's Wives Coalition] meeting . . . was that people were feeling that they really did have to pull together because there's more power in numbers when you get into trouble, despite the image of the independent fisherman.

[In the auxiliary] there have been some tough times with personalities; some personalities are stronger than others. Some of the gals who have been in it for twenty-five years saw it awfully hard to have younger people coming in with new ideas and fresh enthusiasm. [On the flip side,] more [important] than the personality problems was the support we gained from each other. We really formed some indelible bonds that will just go on even . . . [when] the Boldt Decision [is no longer an issue]. Even to the point of people consciously or unconsciously moving to this particular area, which is Richmond Beach. The woods are full of fishermen around here. What's happened is it's a community here of neighbors as well as a community on the dock and on the water. I think that's a compliment, when people want to be in that close proximity to each other; it's a real community.

During the difficulties encountered following the Boldt Decision, Evie found strength through her religion.

[When things hit a low point in the struggle, I had] confidence because of several things. A Bible verse says: "We'll never have more than we can receive." We'll never have any more burdens than we can handle at a certain time. Even though finances are really tough and things looked pretty ugly, I knew that there wasn't any more than we could handle; so that was a major [source] of confidence for me. Also, Christ promises that we'll always have some sort of shelter and some sort of food. I had confidence that Christ was with us.

As a direct result of the Boldt Decision, Evie and Randy have become purse seiners in Kodiak, Alaska.

We wanted to get far enough away from Puget Sound . . . [so that] if there was any residue [from the Boldt Decision] it wouldn't fall into where we were going. Luckily, I believe we went far enough away that so far it doesn't look like it'll touch Kodiak. It can touch Southeast [Alaska since] a judge has now decided that, because the salmon migrate—they go clear to Japan, and they go back and forth in the waters for four years, and then they finally come up the river—[he has now] decided that when those fish are going back and forth, wherever it may be, you can't fish on those fish.[1] They've got to get back up the river for the treaty Indians, and that includes Southeast Alaska, because the fish can migrate up there.

Then . . . [another] . . . reason . . . we decided [to go to Kodiak] was because [Randy's] dad builds Kodiak seiners. His dad could build us a boat a lot cheaper than our having to go buy one off the street. We have fished for a cannery down here, the Washington Fish and Oyster Company, that could finance us for a license. A license [in Alaska] [costs] as much as buying a boat. So, essentially, his dad could finance us for a boat, and the cannery could finance us for a license. And, with just a whole lot of prayer, we could get in fishing up there without much outlay, because we didn't have any money.

Last season [in Kodiak] was good. I didn't fish because you need a crew of four people, and the boat is so small that there is no fo'c'sle and the bunks are just all exposed. To have a woman on board, and then three guys, is just too close quarters. If it was another husband-and-wife sort of setup where there would be two of us, that would be okay. But to have three guys and a gal, it just doesn't work out that well. So, I went up to Kodiak and worked in the cannery. The guys'd come in every three or four weeks . . . and then I'd see them for a couple of days and they'd go back out again. But I think I'll do that again next year.

---

1. U.S. Supreme Court Majority Decision, *United States* v. *Washington,* 1979. See discussion of Boldt Decision under "The Politics of Scarcity," this volume, for further information.

I really loved [working in the cannery]. It was a good job. [My duties were that] I was secretary to the accountant and to the superintendent and to the manager. [I] wrote letters, wrote checks. I wouldn't figure out how much the fisherman had coming. The accountant would say, "Make him a check for thus and such, and split it up into shares of thus and such." And so I'd do that. [I was also the] receptionist.

This was a completely new experience, because I've never worked for a large corporation before. Here was this huge corporation, and money didn't mean a darn thing to them. It didn't bother them to take a half a day off and have a birthday party. I think they spent about $15,000 for a company picnic. My eyes were just ding-dong. The first week I was there, the halibut boats came in. The first check that I wrote was for $187,000 for one boat. I'd never seen so much money in my life; I was just a little kid in a candy store. They have a bunkhouse, and they had a cook that would come around every morning with fresh, hot cinnamon rolls, and the coffee would be perking. They'd have fresh fruit and vegetables, which [were at] a premium up there. Nobody else had fresh fruit, and they'd have pears and grapes.

I learned [while on this job] that the fishermen get ripped. So many games are played. [The cannery] bought the halibut for 85 cents a pound, and they'd sell the halibut for, like, $3 a pound. All they did [to the halibut] was flash-freeze it. It was already gutted; it was frozen whole. They sold it to a middleman, I think. Halibut [ends up] $8 or $9 [a pound]. We have the most outlay—the fishermen—and here [we're] getting 85 cents a pound. The cannery makes at least a . . . 200 percent profit. The grocery store [is] . . . making a whole lot. And the consumer gets ripped. That bothered me. I told them that, and they were really willing [to listen]. They were good people to work for because, except the accountant, they had all been fishermen at one time. They said, "Sure, it's not fair, but that's life."

[I made friends up there and] had lots of good times. It's all a fishing community. The friends I met, though, weren't fishermen. They were from the cannery. I didn't have as much to do with the actual cannery workers as I did with the office staff. A lot of men [and] a lot of Filipinos [worked in the plant]. I was mostly [with the office staff]. They preferred that you didn't mix . . . the office staff with the cannery workers.

[One thing] I didn't like this year, this summer, is that everything was so intense . . . in Kodiak. That's because it was brand new and Randy was extremely nervous having the responsibility of four people, and four families, who were going to live [off] what he did or didn't do. Besides, [there was] the financial responsibility that we'd incurred. All we did was live and breathe fish, and I didn't like that. But I think that was just unique to this year. Next year we can mellow out a little more and not be so intense; take some time and play. We had high hopes of going into the little bays and halibut fishing, and sports fishing, and crab fishing. Well, there wasn't time for one thing.

Evie is not sure what the future will bring, either for herself as a fisherman, or for the commercial fishing industry as a whole.

I don't know [if I will go commercial fishing again in the future]. When we're fishing, there's the intensity [that] is on fishing, and that's all there is. I really enjoyed learning the community of Kodiak and getting involved in the community. I wouldn't learn that if I were fishing. But maybe in future years I would want to go fishing. I don't know.

[If I had children] I think it would be hard for me not to encourage them into fishing. But I hope I could honestly give them other interests besides that. I'm sure that because both Randy and I have the same values of independence and adventure, that they would probably get that from us. But they might be into some area that we are not aware of. I saw fishing as that particular area where you could exercise those values. Maybe our kids would find them in [being] an ambassador in Zambia.

[I see a future for us, fishing in Kodiak], more than down here [in the Puget Sound area]. I hope there's a future. Even during World War II when everybody was supposed to be so into the war, Randy's dad got a deferment from going to the war because he was a fisherman. He fished because he could provide food. I see that as where our future lies. I really see down the road where we're going to have to forget about treaties and all that kind of crap, because we've got hungry people. It could be a problem [if the big companies stepped in].

I wish I had a crystal ball that I could look into. There are a lot of people who are gambling that it's just going to be the big fishermen, the big crab boats, who will be the ones that can get by. We see where fuel is . . . a key issue, where diesel is going to be hard to get. Therefore, I don't want to have a big boat that's going to cost me $100,000 to fuel up every time I go to the dock. I'd rather have a small boat that I can even put a sail on if I had to. Originally [that] is how all those fishing boats started anyway. There's no doubt that we'll be in fishing. I think we just don't know which [way to] . . . adapt, which direction to go. I don't think we'll ever get bigger. I think we're as big as we're going to get. I think we'll just automatically choose not to get big—simply because it's the two of us. We can live on a whole lot less than most other people with families.

Linda Jones *(photo by Sue-Ellen Jacobs)*

Linda Jones

*Linda Jones is a member of the Tulalip Indian tribe. Her grand-
mother, father, uncles, and cousins have all fished. Her family's involvement in fishing has a long history.*

Childhood and Life on the Reservation

My father fishes; my uncles fish; my cousins fish. It would be easiest to tell you who didn't fish. I don't think there are any in my family that don't fish; both sides, maternal and paternal. It's just that; that's it.

I don't know how [my family's involvement in fishing] exactly began. My grandmother, who's in her eighties, told about when she was a child, getting into canoes and going to various parts of Puget Sound to fish. [And later, she would] put up enough fish to last through the winter—smoking them, drying them. My grandfather, who was one of the first tribal judges, did a variety of things to support the family. My grandparents had a real big family. They must have had at least fifteen children, a great number of whom died. Probably about five died in infancy.

[The] first memories I have of fishing [are when I was] about four or five. We'd go down [with a] big, big army tent and camp on the beach. We'd stay down there, even though it was only about five miles from our house, and just stay out. The kids'd play all day while the adults fished. They'd let us sleep in the morning, leave somebody at the camp to take care of us, then come back and get us later on in the morning. That was just a total way of life.

I wasn't raised on the reservation after the age of thirteen. And I don't think I really have any good feel for just how extensive fishing was as a livelihood. There was a great deal of harassment. Indians were for the most part precluded from fishing commercially. It was unlawful for them to fish off the reservation, and it's pretty hard to define what the reservation waters include. That still has not been legally defined. So, a lot of the Indians that did choose to fish had to buy state licenses.

We, as an Indian tribe, are being very much more assertive about the special rights that Indians hold on their reservation and access to fishing on the reservation. We have to explain to people that we are not encouraging our people to trespass, but they do have certain rights. I think that by and large [Indians and non-Indians] don't differ that much [as far as their involvement in fishing is concerned]. If it's anything, some of [the non-Indians] are a little more racist because they feel as though their rights are being encroached upon by Indians asserting their fishing rights.

When Linda was seven years old, her parents separated and she was placed in a non-Indian foster home on the reservation. At age thirteen, she moved to a non-Indian foster family off the reservation. Because of these experiences, she began to make distinctions between Indians and non-Indians, and between reservation and nonreservation life.

I was taken from my parents at age seven and put in a non-Indian foster home on the reservation. I was terribly unhappy. It was a really severe cultural change for me. I was formerly living with relatives my age. We all slept together and shared clothes. We were generally pretty happy. Once I moved into the non-Indian foster home, I was no longer allowed to associate [with Indians]. [My foster parents] bought me all new clothes that I wasn't really comfortable in, cut my hair off, and gave me a perm. I remember being terribly lonely. It was a real trauma for me to sit down with the family and eat for the first time. I was terribly self-conscious and totally out of my own element. It took me several months before I adjusted. At that time, I was told that I had no choice—that I had to be where I was.

My mother came and visited me sporadically. I finally realized, as a result of the real restrictions on me by the time I was age twelve, that I no longer wanted to stay there. It was not only the fact that I couldn't have my own family visit me, but the fact that I was not allowed to leave the yard. I was not allowed to have friends over after school. I was not allowed to go to school games or school functions [or do] all the things that my friends were doing. [And I was] discouraged [from] having contact with any of the Indians on the reservation. There was no really good reason given for that, except that I just couldn't. So, things started building up. I was never really one of their own children. There was always a distinction, and it was not a very loving atmosphere. Real differences came about when I reached the rebellious age of thirteen and ran away.

If I had been allowed to have contact with my family, I might have been able to stick it out. But as it was, it was just intolerable. [My Indian family would] go by and wave at me, and . . . if my foster parents saw me waving, they'd call me into the house and make me stay in the house. It was really, really bizarre. I think that instilled in me a shame of sorts that I was an Indian, that I actually belonged to the race and was not on a par with everyone else. So I ran away and spent the summer with relatives. [I told them] that I didn't want to go back, and they didn't make me go back.

After spending a summer with my relatives, though, it just wasn't possible for me to live with them. So, I was put in another non-Indian foster home, in Marysville, which was much better. They had girls my own age. They made no distinctions between me and them. It was a lot more normal family atmosphere, which is essentially what I was really starved for. They also allowed me to visit my family on weekends and go home to the reservation. My first experiences weren't pleasant, though, going back. Just the fact that I'd lived a different life-style made me unacceptable to a lot of Indians on the reservation. I just went back for visits before I graduated from high school. Those [visits] were not very frequent. I generally got caught up in my new family and did family things with them. It was great.

After I turned eighteen, I was so conscious of the fact that I was being state-supported. My foster parents told me I could stay. They were still getting checks for me. I felt guilty about that, for some reason, and I really re-

sented it. So [after graduation from high school], I moved out on my own
and went to school. Then began a gradual return to the reservation, visiting
my parents and getting to know my family again. It took me a couple of years
to really be accepted. There were a lot of people [who] thought that I thought
I was better than they were simply because I was raised in a non-Indian home.
And some people still are that way. I just didn't fit in the traditional niche
of a woman on the reservation.

When I was very young [and] living with my parents, the reservation was
no different than any other place. I didn't realize then it was a reservation.
And I think that probably even until I was about fifteen, I never really thought
about the reservation. It became most distinctive, I think, with the advent
of the federal court rulings affecting Indians and special rights held by Indians on reservations. It's just now so much a part of everyday life that it's hard
for me to tell somebody that doesn't know anything about it. Most people
come out to the reservation, and they'll stop at a store and they'll ask, "Could
you tell me where the Indian reservation is, please?" And they're already three
and a half miles into it.

[The percentage of non-Indians and Indians living on the reservation] is
about a three-to-one ratio: more non-Indians than Indians. All the prime
waterfront property is leased to non-Indian families. One of the things that
differs is that non-Indians leasing Indian land don't pay property tax. They
have to pay a personal property tax, but they don't pay a [property] tax *per
se*. I think there may be a small savings there for some of the people, plus it's
just a really pretty place to live. We're so close to Seattle that there has always been a lot of different people on the reservation, even when I was young.
[My] first non-Indian foster home . . . was on the reservation. There was
already that much settled by non-Indians and by Indians from other reservations.

[Social distinctions between Indians and non-Indians on the reservation]
didn't really become apparent to me until after I was taken from my family
and put in a foster home. Then there was a great distinction, primarily because I wasn't allowed to associate with the Indians and wasn't allowed to
visit my family anymore. [It] was really unpleasant. . . . I was simply told
that only my mother could visit, because my foster family didn't want Indians in the yard. They didn't want the neighbors to see Indians in the yard. I
really couldn't understand that, because I was an Indian living under their
roof. It was really odd.

There are probably about five main Indian families [on the reservation].
Even some of those are interrelated. You've got [five] dominant family names.
I'm related to [four of these]. There's so much interrelation, with the rare
exception of when a non-Indian marries into the Indian families. [Pressure
for Indians to marry Indians has] gotten fairly relaxed, I'd say, in the last five
years or so. It would be nice [for me to marry an Indian], since I'm almost a

full-blooded Indian. There aren't many of us, and the more we can help the Indian population along, the better. But it certainly is not my prerequisite. I used to feel guilty for not thinking of marrying an Indian. I thought, "God, I really should marry an Indian." There are Indians that definitely feel as though they would not consider marrying anybody but another Indian. I visited another reservation east of the mountains where they simply do not allow non-Indians to go up into the mountains. They have guards posted there, and unless you're an Indian and can provide them with a tribal number, you don't go in. They're very prejudiced, and they attribute a lot of people's bad personality traits or habits to the fact that they must have some non-Indian blood. Very prejudiced. And you know what's worse? I enjoyed my exposure to it.

Fisheries Management

After graduating from high school, Linda went to college and began working as a legal secretary in Seattle to support herself. She became a paralegal with the Washington State Reformatory and then moved into fisheries politics and management.

Just prior to going to work for the tribe, which I did a year ago in September [1979, I] worked for the Northwest Indian Fisheries Commission for about three years. I started out as the assistant coordinator for the Point Elliott Treaty Area. My primary function was to coordinate the fisheries management program of the nine Point Elliott Tribes. We started out with drafting a constitution, agreeing on which areas we would jointly regulate, who would have primary regulation of other areas, and generally making sure that [the nine tribes'] regulations and fisheries practices didn't conflict with one another. The highlight of the whole month was a staff meeting when we would all travel to Olympia and meet together. I think we had really good quality people, because we weren't in it to make money. We weren't in it to glorify anybody's name or make a name for ourselves. We were there to do a job, and we really believed in it and were just very dedicated. We put in long, long hours.

Prior to that I worked in law offices. I almost became a lawyer, but I felt I could serve better as an advocate. What a lot of people feel [is] that lawyers are too conservative and too worried about the consequences of possible actions to really be effective.

I only got involved in fisheries because I was looking for a new line of work. It was the most likely area for me to get into, since my family has fished for years and years and years. After the Boldt Decision it seemed to be an area that needed a lot of attention. There's so much public opposition to a decision that gives Indians half of the harvestable fish when they comprise no more than 1 to 2 percent of the total state population.

I'm the Director of Fisheries for the Tulalip Tribes, and that includes a variety of responsibilities. It varies. Just the day-to-day management activities include promulgation of regulations of the fishery; supervision of the staff (that generally numbers around ten—biologists, technicians, secretaries, and anthropologists); generally being responsible for the operation of the entire fisheries program; and reporting to the tribal governing body. I'm the second fisheries director the tribe's ever had. [The job is] what I am making it. Fortunately, I've been given a lot of latitude, so that I'm not restricted to just fisheries. I don't have to just do day-to-day management work, or just represent the tribe at meetings. [I can] get involved in local politics, elections, that sort of thing. I definitely have to get approval from my higher-ups [for such activities], and I go to them for general guidance if I'm not sure that what I'm about to embark on is acceptable. Other than that, I do a lot of public relations on behalf of the tribe, including a lot of speaking engagements and participation in a number of activities relative to treaty-rights protection.

[In a typical day] there's generally a lot of meetings. A typical day will start in the office with a series of phone calls to different people for a variety of reasons, coordinating staff activities, and running meetings. Meetings are commonplace; they occur almost every week. We recently concluded a series of allocation meetings with the south Puget Sound tribes. The purpose of that was to determine what percentage of fish destined for south Puget Sound we would intercept, and what numbers of fish we would guarantee them for their harvest. A tremendous amount of delegation of authority occurs in my office. Even though I am the director of the program, there are people that were there before I was that know what their jobs are. [They] just do them and function well without my daily meeting with them, my supervision of them.

I also represent the tribe politically within the Boldt case area. [I work with] all the tribes affected by the Boldt Decision. We're all organized into treaty areas, and ours, Point Elliott, is the largest and consists of nine tribes. I sit as the [nine] tribes' delegate to the [Point Elliott] Treaty Council and chair [that] council. I [also] act as alternate commissioner to the Northwest Indian Fisheries Commission for the . . . treaty council. When the commissioner can't make it, I represent the nine tribes in giving direction and making policy decisions on behalf of the tribes. The Northwest Indian Fisheries Commission is the coordinating body . . . created after the Boldt Decision to implement the decision and generally upgrade the quality of fisheries management.

[Compared with the job of the main fisheries manager for the state Department of Fisheries], mine is much more localized. The state of Washing-

ton houses their fisheries director in Olympia with very little personal contact with the fishermen—thus, a lot less pressure, I'm sure. It's very unusual
[for a woman to be involved in this work at the administrative level]. I think
in all the Puget Sound, I'm one of two women. I'm the youngest one in the
Puget Sound area. I've recently begun working with Congress and members
of their staff to ensure that the federal court decisions affecting Indian treaty
rights are not adversely affected. I recently spent a little over a month [1980]
in Washington, D.C., working with Congress and will probably return after
the election to see if the bill does pass in the Legislature.

[I got to my present position because] I think that people were impressed
by my background, by the fact that I've worked in law offices and have had
as much education as I have. I probably [have] more education than a good
number of the people on the reservation. I'd also done volunteer work for
the tribe. I used to spend a couple of nights a week [on the reservation] ad-
vising people on [what] legal matters I could, and referring them to appropriate community agencies or whatever. Consequently, after working in that
capacity and then after working for the Northwest Indian Fisheries Commission for as long as I did, I was the logical choice. But I think that [another]
reason I had a fairly good working reputation at the commission was [because
of] the fact that there was a free exchange of information necessary to make
a decision—not just what people wanted them to know. So, I was always
very candid, very straightforward with people.

I enjoy [my job]. I love it. I get to do such a variety of activities. I'm involved in so many different things, and it's a challenge. It seems as though
once I feel I've gotten something well enough under control, I can turn it
over to my staff and then they'll follow it. [This allows] me to go on to something new and get that integrated into the fisheries program.

Managing Job Pressures

*Linda moved back to the reservation when she began working with the Northwest
Indian Fisheries Commission. This created problems for her, when she became the
Tulalip fisheries director.*

After a couple of years [of working for the tribe], though, it wasn't just a
job. It was more a way of life. It's hard to define where the job ended and
my personal life began, because it was so intermixed.

[My] job [as director] is hard to separate from my personal life, and the job
has become a way of life. I recently made a real conscious effort to separate
my personal life and my work life. [When a person is] working and living on
the reservation and having family that is so actively involved in fishing, there
begins to be no separation of the person that, say, goes to the office from
8:00 to 4:30 and the person that goes to visit family on Sunday afternoon.

Because the first thing people want to talk about is fishing. The first thing my cousins want to know is when [our fishing area's] going to open again. I just sort of lost a lot of the camaraderie I had with my family [and] my friends. And I got real tired of it, because I needed to have some time for myself. I needed to get away from it, because there is so much pressure. I just needed to socialize like I used to do. So, I moved to north Seattle [to] a neighborhood where a lot of my friends live, because I found that I was spending a lot of time in Seattle anyway, just to be able to have some sort of release from the day-to-day pressures that build up. I find that I'm a lot more relaxed and a lot happier. I tend to be able to deal with things a lot easier, because I have a forty-minute commute each day that allows me to think and plan a day's activities, or, on the way home, to unwind and let the tensions sort of fade away.

[In Seattle] . . . people don't bug me about the role as fisheries director. The fact that I'm single makes the need greater to get alone. I face an awful lot of pressure whenever I socialize with people in the community I work in. It invariably ends up that I am at a party and somebody will want to talk fish. Because when they see me, they don't see Linda Jones; they see the person that runs the fisheries program. They want to know what's going on and why. People that won't bother to stop by the office and find out what's going on seem to think you're fair game if you're out and about, partying around.

Although Linda relieved a lot of the stress of living and working on the reservation by moving to north Seattle, she must still cope with daily pressures of the job, such as being caught between tribal leadership and the fishermen, dealing with other tribes, and working with politicians.

[In terms of tribal leadership] there's always a swing. You'll go through phases where one family will have a lot of involved members that are knowledgeable in fisheries. [They] are viewed sort of as being a real credit to the tribe and the good guys in all this. And then eventually they screw up . . . or they just really disillusion the tribe or misrepresent the tribe, and [they] don't get reelected. Presently on our board there is a brother and sister and almost everybody's cousins one way or another.

I have seen instances where family pressure brought on [members of] the tribal council forces [those council members] to put pressure on me—say, for instance, to open a fishery or to provide off-season work for some of the fishermen. I really feel I'm fortunate. I don't feel a whole lot of family pressure. My family's really supportive and they've always taken a defensive position for me when things start getting hot. My father has never put pressure on me to do anything in my job that would directly benefit him or that would in any way compromise my integrity.

In the area of fisheries, Indians having their own management authority is relatively new. It's not something we've done for decades, even though we did practice it, in our own way, years and years ago. I think basically the tribal staff has given a very vague outline, a skeleton, because nobody really knows what the hell to do. Consequently, they [offer] some basic principles [from which] you're supposed to interpret what tribal policy is, what Indian policy is. If everything goes well, great; but if something goes wrong, then you know that generally the [fisheries] staff will bear the brunt of it. That's probably true in just about any job.

The most immediate pressures come from the fishermen that feel that Indians have a right to harvest 50 percent of the salmon [and say] "Let's go to it." [They] don't realize, to nearly the extent necessary, that along with the right to half of the harvestable salmon comes the responsibility to manage the salmon resource [in order] to assure that the salmon runs are perpetuated. There's always going to be the pressure from the fishermen for a longer season, for more fish [and] for less sharing with other tribes.

[Sometimes] . . . it gets me down, having to contend with the ego of the man that is the highliner in the fishing fleet that thinks he knows more about the fish runs than our biologist does. He knows more about this tribe than I ever will, and the attitudes are pretty blatant. Consequently, I don't give in easily . . . and [it] makes for some unpleasant conversations sometimes. [There are also] . . . bullshit stories that go on in the circle of fishermen about things that the fisheries department has done, and [I get] tired of having to file reports for every little incident. I'm being kept so busy writing reports on why the docks security personnel didn't walk down the dock every hour, and why the patrolmen did this, and so on, that I can't concentrate on the things that are really important to my job.

Probably secondary to [pressure from local fishermen] would be political pressure from other tribes, because we realize that the fine tuning to salmon management has to occur now. We're realizing more and more that salmon passing through our fishing areas are destined for other river systems and we've got to realize their needs and develop our harvest management methods.

Politicians [are another source of pressure]. We're very concerned about any attempt to legislate away any portion of the Indian treaty fishing rights. [We] are really sensitive to that [so we] are making a conscious effort to work closely with Congress to ensure that Indian rights are protected.

In order to make any positive contributions in Indian fishing [and] Indian treaty rights, you've got to care. And after so many negative experiences, I don't think that one can be expected to always be positive. After being told that your fisheries management program is a total failure, and that you've lost fishermen thousands and thousands of dollars, you don't really feel very good about yourself. That's not good for your ego, and with never a chance to really deal with those accusations on a one-to-one level, you just start wondering why you're in the field you're in, why you're doing what you're doing. And why should you care anymore? Why should I care enough about the tribe to hang in there [in the face of sustained criticism]? The answer to that is my family, because they've been so supportive and they've allowed me to maintain some incentive to stay in there.

Sometimes I feel hypocritical—I haven't lived as an Indian. It scares me. I hesitate when I make a decision on behalf of Indians. I [sometimes] feel like a token Indian—that I don't have the cultural background I should have. [Also] I'm afraid of being referred to as an urban Indian. There are different classes of Indians, just as there are in any other society. There is a certain distaste or scorn for urban Indians, because they are literally people with no land base. Urban Indians live in the city, even though they might be from a reservation and they've just relocated. It's almost as though [urban Indians] don't want to be associated, [as though] there's something there that they don't like and they want to start all over. I don't think that the general conception in the Indian communities is really all that favorable. So, I find myself thinking [about] making the decision that is in their best interest, or what I think is best for them. It's really hard.

The [best part of my job] is probably the good feeling you get when you're out on the boat with one of the fishermen. Then you realize how much they enjoy their way of life and how your efforts to ensure that that continues [matters]; that there's actually fish for them to harvest . . . [makes it] . . . worthwhile. You just have an underlying sense of doing what's right. I've never, ever felt that the line of work I'm in is as bad as a lot of people would portray it.

The bad [part] is overcoming the prejudice in the Indian communities [of my being] a young woman in the fisheries management field. Its fair to say, I think, that Indians are fairly sexist and that there's always been traditionally a great deal of respect for the tribal elders, for the older people. Consequently, having a young person come in and give orders and to decide when and where you'll fish and when you won't is real hard for some of the older guys to swallow.

Women and Fisheries

Linda has had to deal with personal and professional pressures deriving from the fact that she is a young single woman who is also a manager. She admires women who are active in the industry, noting obstacles and ways to overcome these as well as prejudices against women.

Recently, I've felt the frustrations of being a woman in this business. I don't know if it's the time of year; I know I'm not the only fisheries manager going through . . . problems . . . with people's dissatisfaction with the decisions made during the season. But it seems to me, anyway, to be tougher to be a woman [in fisheries management] because it's so male dominated. There are always a score of men that just have a really hard time relating to women [in positions of authority]. There are men who feel threatened by competent women in a field, and [who] do their best to discredit them or slight their expertise in the area. They don't think of you as being a competent individual first. They think of you as being a woman, and therefore a little weaker, a little more emotional, and probably not as capable as they are. [And there's] always the little "good ol' boys' club" and the jokes that you hear about new women around, after you've sort of become part of the accepted circle. A lot of patronizing occurs, and it's almost as though, even though [women] were elected officials of [some] Indian tribes, there seem to be different people that would rather we not be involved in the decision-making process. [They] have a real unpleasant attitude when it comes to dealing with [women].

Another Indian woman with a position similar to mine is accorded a certain amount of respect because she's married and she has a family. She gets kidded somewhat, but I think probably there's not nearly the speculation about what she does when she's not on the job as there is about me. One of the most difficult things are the sexist jokes that generally get directed at a single woman in an all-male crowd. I don't find that pleasant and, unfortunately, I don't have the ability to take that sort of thing lightly. I get offended sometimes. I think that if I were to be as crude as they are to me, and reverse it, that they would be shocked and they wouldn't like it. Obviously, I'm not that sort of person, but I wish that they would think of that when they're directing some of their crude remarks at me.

[In] Indian government, Indian men are dominant . . . even though there are a lot of families where the women are the actual heads of the households. I think that the tribal councils are predominantly men, and the women that do make it on the council are generally old family. [The Tulalip] never had a chairwoman or chairperson on the tribal council. It's always a chairman. And that's just the way it's always been.

In the particular field that I'm in there don't seem to be too many women involved. Probably what you see most is young women emerging from various tribes to represent the tribes.

When I was hired [as director], I was the board's first choice. But they chose an assistant for me, to more or less take care of some of the technical day-to-day things, because they recognized that most of my exposure and experience was in the political arena, and not necessarily in any of the technical management and enhancement aspects. So, naturally, they chose a man. The support staff I have is one that has been with the fisheries program since it started. One central individual, until recently, was no more than a secretary. I have since promoted her to administrative assistant and have attempted to get her to assume much more responsibility. [I hope] she'll be able to assume more of the leadership role in the office when I leave, because I know that the tendency won't be to replace me with another woman. She's a gold mine of information because she's been there so long. It's the sort of thing you can only accumulate after being exposed to it year after year after year.

One of the toughest things about being a woman is [that] people don't think you know anything . . . especially in fisheries. I encountered that in my first month on the job at Tulalip—people getting irate and actually in fishermen's meetings saying, "What do you know about it? All you do is sit behind a desk and give orders. You've never been out here fishing. You're just going to screw everything up." A lot of [men] felt as though a man had to have that job, preferably someone who had fished before so [he] could identify with the real interests of the people. Fortunately, there were two fishermen that applied for [the director's position when I did]. [The board] felt . . . they had to have somebody with a broader base than just the actual fishing experience, because fisheries management consists of so much more than just deciding when you want to go fishing. A whole lot of politics [is] involved, and [Indian] men have actually run things for so long that it's really foreign to have a woman in there.

[The woman I mentioned earlier and I are close] whenever we're out in a meeting together. It's all men, say, for instance [at the] United States/Canada [International Pacific Salmon Fisheries Commission meeting]. We weren't ever very far apart [from each other] there. It's just because it gets overwhelming sometimes, and it's nice having another woman there. [When working with another woman] there's almost an understanding between the two of you.

I think [other women, seeing women in fisheries for the first time, are] surprised. They watch you like a hawk to see if you know what you're talking about, to see if you know what you're doing, or if you're just a figurehead of sorts. Then you tend to get into one-on-one conversations [and] generally come away allies. I think I work well with other women, and I work well with men—most men, not all men. I think that women might be a little more diplomatic [than men] and have a little less ego and . . . be better negotiators. They're less threatening to deal with than other men are and probably listen a little better. I could be wrong, but I think that. Sometimes, dealing with other women, I will . . . agree with them, but my bottom line has already been established for me and I can't go below that, even though I think that what they're asking is not unreasonable. I've been accused of trying to give away all our fish, anyway. I haven't had a whole lot of contact with many non-Indian women in my job.

I haven't really thought about feminism in the industry. I am not closely acquainted with feminism at all. I think it's just women that have worked hard and that have [a] vested interest in fishing. Say their family's been in it, or they've been raised around it and they've worked their asses off, literally, to get where they are. None of the women I know personally are feminist at all. The Indian women I know don't get into feminism, perhaps because the Indian society is a little different. My grandmother is the head of our family. She's certainly not a feminist, but she gives the orders and the people obey out of respect. There are a lot of families that are maternalistically inclined . . . where the mother is the head of the family. We identify less with what I've heard termed "role playing" than, maybe, the non-Indian society; but, it's just not this all-encompassing feeling.

We don't have any women technicians. It's just generally accepted that women aren't sturdy enough, I guess, to work with the men out in the elements. In the fishing fleet, I'd say we probably have a dozen or more women. They're gillnetting or beach seining or have their own set nets. The total fleet is about thirty-five to forty marine gillnet boats. [There are] about ten beach seine boats and about fifteen small skiffs. So, it's not a real large percentage, but there are more. There are two women that just this year [are] starting to get into gillnetting and now own a big thirty-two foot boat by themselves. It usually starts with women going out with their husbands or their boyfriends or, in a couple of instances, with their fathers. Because they've found nothing else they want to do and it's so much a part of *their* life, that's what they choose to do to earn money. There's also a lot of conveniences that weren't available a long time ago, like hydraulic gear. Some of the equipment is much more sophisticated, which doesn't necessarily mean you have to have brute strength to fish. I think that [fishing is] not easy to get into and it's not easy to stay in. I would applaud other women that get into it and can be strong-willed enough to take it. A lot of women are also involved in the off-season in readying the gear for fishing and generally being real support persons in term of repairs on the boat and getting ready to go for the coming fishing season.

I'd say from personal experience that women have been able to accomplish things that men couldn't. There are times when women can make everybody important, and that's really a good quality because everybody needs to feel like they have some say in what's going on. [It's good] if you can pool all of the ideas of people like that and come out with one end product that is satisfactory to everybody.

Reflections on the Past and Thoughts on the Future

Linda feels that her experiences on the job have made her a stronger person. She also believes that there is a future for women in politics and management in Indian fishing, if they are willing to be tough and persevere.

In the last seven years, I've grown emotionally. I've discovered my abilities and learned to appreciate myself. I didn't used to have a very good opinion of myself, or maybe I doubted my abilities to provide a living for myself, or even [doubted] that I liked myself very much. I was terribly insecure after having been through two foster homes, not really realizing who I was or what I was going to do for the rest of my life. I felt this responsibility to do good, to do something to support myself, especially. I can't identify the exact time when I stopped making decisions because I thought it was what I should do, but started doing things that I wanted to do and felt really proud of myself. [But, I found that] I felt like I was in charge for the first time. [I] actually found that it made my dealings with other people a lot easier. I was able to be more pleasant and just a better person.

Right now [there are no limits to how powerful . . . or how influential a woman can be in fisheries] . . . if [she] wants to hang in there that long. I think that the tendency for anyone in northwest fisheries is to burn out after some years—to want to find another job or just take a break for a while. [I am optimistic about women in fisheries in the next decade] only inasmuch as they're willing to just last. I think that there are going to be more and more women becoming involved with fisheries, primarily because a lot of the men seem to get involved in the active pursuit of commercial fishing, whereas more women are getting more education or becoming more acceptable as managers. [But, I would caution] anyone planning to get into the political arena of the northwest fisheries [to] be prepared to be really, really tough, to hang in there and not become easily discouraged by a lot of the sexist attitudes displayed by men. And [as a woman] you've got to work twice as hard to prove that you are really capable and competent. There's always a bit of skepticism, I think.

The first thing you've got to do at all times is to maintain a firm belief that what you're doing is right. You've got to be a staunch believer. In the field of Indian fishing rights, you have to be able to weather a lot of criticism, a lot of hostile people, especially on public speaking engagements. There are a lot of problems in the fishery that are attributed to the Boldt Decision [and that attribution] is just total b.s.

When I first started in this field, I spent probably the first year and a half to two years sitting in on meetings not saying anything. I really didn't know what the heck was going on. I was not qualified to make any statements. You have to really believe in yourself; you have to make sure you know what you're talking about before you do talk. There's nothing wrong with saying "I don't know" or not saying anything. I think I've always been fortunate in having people set good examples for me in the development of my people-skills, working with people. [The] people [I worked with] helped me establish my own working patterns, my ideals, my goals. They were good-hearted people that worked hard and were dedicated and cared enough to try to share what they knew with me to get me off on the right foot. I appreciate it.

I feel really fortunate and honored to have accomplished what I have at my age. It takes a lot of other women longer. I feel as though I am doing a service for my family in the job that I'm doing, and even though I [have faced] a lot of undue pressure from [various] people, I can go to [my family] for moral support. I think there are even some men who don't enjoy the status that I have. I feel as though it has opened a lot of doors for me. It's just really rewarding.

Gladys Olsen

Gladys Olsen

*Gladys Olsen is a native of Alaska, where she was born in 1910.
She has spent the greater part of her life on islands around Kodiak,
Alaska. In 1969 she moved to Poulsbo, Washington.*

Alaskan Heritage

I'm part Russian, part Aleut. My grandfather Chichinoff was born in Fort Ross on the Russian River in California. Both [sets of grandparents] had the Russian and the Aleut. My father and mother were the same. They were born and raised in Alaska. My father's side was born around Afognak. Afognak Island, that's where I was born and raised. It's about . . . thirty-five miles away from Kodiak. At that time [the population was] around three hundred to five hundred. It all depended. Afognak village is—was—out in the open. The Gulf of Alaska would roll right in whenever we had the northeasterly storms, and then the northerly storms were bad, too. The tidal wave of that earthquake we had in '64 . . . destroyed the village completely. It had to be evacuated. They started another village, which they called Port Lions, and that's on Kodiak Island.

[If we needed emergency medical help] we had to go by boat to Kodiak. If there was a doctor there [we got help], and sometimes there wasn't any. A lot of times we had to depend on the old-fashioned methods of medicine that . . . the Alaskans had. I have lived a sheltered life and . . . I really do not know the culture and the habits that our people had before. A lot of the people in my time were like that. Seems like we weren't told, and by then it began to disappear. There was a little bit of superstition. But by that time our generation was getting so we didn't believe in that superstition. [I completed] eighth grade. That was [common] in . . . Afognak village. In Kodiak it was different. They had high schools there.

I had a happy childhood, a very secure life. My father was a carpenter and he worked at canneries, usually building skiffs that the canneries had needed for their fishermen to go out to fish in. My mother was just a housewife [and] raised a large family. [There were] thirteen [children]. They lived for quite a while before some of them began to pass away. All my brothers are gone; I had five brothers. There were eight girls. I had a brother who was a machinist. One had to go away because he wasn't well and he was put in a sanitarium. The third one, he also worked in the canneries and most of the time around machinery. The fourth one . . . he was also a machinist . . . in a cannery and other places. The youngest one . . . loved to cook. He was in the army . . . a while and did quite well. Then he got sick and passed away, so there wasn't much to his life. My sisters . . . all went to work, either in the canneries, or housekeeping, or something like that.

I've worked in a cannery before I married. I went with my mother. [She] had to go because my father worked at that cannery, a clam cannery in Kukak Bay. [It took] six hours to go across by boat [from Afognak]. It took us three hours to cross the Shelikof Straits alone, by boat. [So] the whole family had to go to that place. It was all right for the children to be there. It was all right for the wives to be there. I was only about fourteen, fifteen years old. Yes [it was common for women to work at the cannery]. [I was] clipping clams—clipping and cleaning them, getting them ready for canning. I liked it. [It wasn't boring], not for me. If it [was boring for other people], they didn't complain much. They knew they had to earn their money. A lot of those people had come from Westport and . . . Ocean Shores [in Washington], because there were clams down here on the beaches and the water [so they knew about clamming]. [The money I earned was] very, very small. I got so I could do six hundred pounds in a day. I enjoyed it. [My mother] was only there that one year. My father didn't go back anymore after that. After that, just [two] of us girls used to go across [and] my brothers, three of them. Most of the women that came up there were married. [They] just worked in the cannery while their husbands were out clam-digging.

Marriage and Beginning a Family

When Gladys was nineteen she met her future husband, Hans Olsen, in her home village of Afognak.

[I was married at] twenty-two. I guess kind of oldish [for my generation]. Hans . . . was a Dane from Denmark. He sailed—before he came to America—around the world practically, except the Orient. He was . . . twelve years older than I was. But our life was beautiful. Age doesn't matter, really. This can be proven. My husband passed away [in 1975].

My father knew him, and my family, most of them. My father worked in the canneries, like I said, as a carpenter. Hans had worked the tenders and, of course, people that work in a cannery, they mingle together. They know everyone. Hans decided to move to this village, Afognak, and I had cousins who also worked at that cannery and who knew him. Well, of course, they talked about little Hans, and I wanted to meet him *so* bad. We were taking a walk, "Let's go and see Hans." "Well," I says, "I do like to meet him." When we got there, there were two of them baching, but they had company when we went in there. They were all shy, but they were laughing at something. Here Hans was making a pie. It happened that he would put too much shortening, and then he'd put more flour because it'd be all shortening . . . and it'd get too stiff. That's what they were ashamed of. He had it in the oven, and they were afraid to peek because us women would know what had happened. So that's how I met Hans.

We were engaged for two and a half years, I guess. No money. Waiting until both of us could have a little bit more money to get married. But then in the end—why, we got married and . . . didn't have much money. I worked in a cannery the first year [we were engaged] and . . . he worked on a tender. Those days they had fish traps; he helped put that up. He worked on a pile driver, and then in the fall—why, they pulled [the piles] out. Finally it got so that we decided, money or no money, we're going to get married, which was good. We were able to make good. It took us a while before we started really putting money away.

After I married I didn't work. I stayed home and took care of a family. After two years I had my first child. We had a nice family, a very nice family—seven sons and one daughter. She was the last one. I had midwives [for their births. And] I had two doctors—one with the third one, and then the last one, the daughter, was born in hospital. I had no problems. Of course, [Hans] wanted a daughter to start with, but we had to have the boys to start with . . . all together. I didn't mind; he didn't mind. He was the nicest man, the nicest father. But when our daughter was born—I knew he wanted one from the beginning—I sent him a telegram. It was from her, see—"Arrived 10:30 A.M. Mother doing fine. Anxious to meet my seven brothers." Of course, by then we had the name. He folded that, kept it in his wallet for years and years, until it tore in the creases.

A few years after their marriage, Hans went to work in a cannery on the west side of Kodiak Island. There was accommodation there for the single women who worked in the cannery, but nothing for a family. Gladys and her two young sons went to live on nearby Dry Spruce Island.

I stayed on an island . . . that was called Dry Spruce Island . . . for six years with just two sons to start with . . . and they were little. I had the third one there. We had a little cabin. If you had to row [from the cannery to the island] maybe it'd take you twenty minutes to half an hour, but then with an outboard motor, say, fifteen minutes.

Most of the time I was there all by myself [with] the children, and Hans would come home . . . whenever he could. Hans worked at the cannery, either the general work or working in the spring when they start getting prepared. Like they take care of the nets, boats, and things like that. He worked in the wintertime there, too, if there was a winter watchman. He worked with this winter watchman, seeing to it that things for next year were prepared and in good shape. They always had something to do. In the . . . salmon . . . season, when the fish was at its peak, why, he never came home—maybe just for two or three hours, or overnight. That's all. He was home more often during the winter, at least once a week.

The winter watchman's wife, when I was on the island . . . would come over and visit with me. While the men'd be working, we'd go back and forth. Not far away, though, there was another island [where] I had a sister and her husband and children. Her husband had been raising foxes, at the time when fox tails were good money. But then it got so they didn't cost much anymore. So he quit. Then they moved away.

Yes [it was lonely for me]. I was young. But then I had the children; I was kept busy. As a young girl I dreamed of a little bit more—maybe go see a movie, or have more people coming in to visit with me, especially when I stayed on the island. But most of the time [I] never thought of it.

I never, never suffered. We had the food from the beach . . . clams . . . [and] crab. He'd get salmon we'd put away for the winter. We also had cod-fish, ducks. We didn't have no money, but then we had food. I always had a garden. Up in Alaska you cannot grow as much as you could down here [in Washington]. But then [you could grow] the essentials, like the potatoes, the carrots, rutabaga and cabbage, beets, things like that. The cannery had stores [so Hans] would bring home what I needed.

Gladys and her sons were still living on the island when World War II began. Because the island was by the main shipping channel, she witnessed the movements of convoys of commercial boats protected by U.S. Navy destroyers. It made her very uneasy.

[The convoys] all had to go past the island. Lots of times I used to watch them, and maybe they'd have to stop on account of the tide, and then they'd have their signals. We used to watch that at night, the children and I. We had to have darkness—blinds, dark blinds at all times during that time. [We heard the news] by radio . . . what they were able to give out. You see, everything had to be secret. I was on the island when the [Japanese] bombed Unalaska, Dutch Harbor, and Kiska Island—and Attu. When I heard about that I was afraid, because I was in the main channel there, the main course . . . and here I am all by myself. So, my husband and I talked it over. We found a place in the middle of the island where the trees and branches [had] overhanging moss. I always kept a box handy of everything—food, blankets, clothes. Always the children would be ready to dress.

This one day we were having lunch, and that was about the time when the [Japanese] were in Kiska. The ships were going up there, so that kind of worried me. We were eating lunch at the window. Jutting out . . . there was a point, and it kind of curved into . . . a little cove. All of a sudden a big ship came around that point and just stood there. And I thought, "Oh, my God! There's one of the Jap ships, and I'm here all by myself!" I started getting the children all ready. They got their boots on, their heavy macki-naws, and I got myself all ready. It was foggy, too, and we were so afraid. I was afraid. Then the fog lifted a little bit, and there was the American flag. Oh, how happy! The most beautiful sight. The American flag.

Then, I guess, the cannery had radios that they could listen to what the boats were talking about. So they knew. Pretty soon the tide was high enough so that the cannery tenders could come through the narrows. See, the island would dry up into the mainland when the tide was out. But this was high tide, so they went by and . . . Hans saw me outside with the children. He knew everything was safe, and they waved as they went on to the boats. Here this boat [had] got lost from the convoy and got stuck on that reef. Later a minesweeper came and . . . was able to get closer to the ship. They helped unload the canned salmon from the Bering Sea and put it on another ship. That way they were able to get the boat out of there when the tide came in. But that was one of my most interesting times that I had with the children on the island. And to see that beautiful American flag!

Life as a Fisherman's Wife in Afognak

After six years on Dry Spruce Island, Gladys's husband decided to buy a boat and go salmon fishing. Soon afterwards, the two older sons, Thorven and Hans, were ready for school, and so the family moved to Afognak village where Gladys had spent her childhood. In Afognak, Gladys was back among relatives and friends who provided company and practical help. She saw less of Hans, however, than she had when he was at the cannery. He soon started to fish for crab as well as salmon.

The last year . . . that I was there [on Dry Spruce Island], he decided to buy a boat and go salmon fishing . . . and he was a salmon fisherman ever since. Then my children had to go to school, so I had to move away. But to start with, two of my children were old enough to go to school, yet we kept them so that I'd still be close to my husband. The first child, he was eight years old when he went to school. But I taught him . . . and so did Hans. He helped. He had a lot of patience. [It wasn't common], but because I was on the island then, we had to. The first one . . . when he started school [in Afognak] was half a year in the first grade, the second half in the second grade, and then he was promoted to the third grade. So our teaching did a little good.

After my husband had gone out fishing, I was very much alone with my family. I stayed home and took care of the family. Saw to it that they were raised according to what we thought was right and proper, which we did quite well, I thought. We all felt it was our duty to raise a family, and that's just what I did. I stayed home, took care of them, saw they went to school. After . . . [Hans] started crab fishing, well, then he was gone in the wintertime. He had his own boat . . . and his own gear. The first little boat . . . maybe it was [a] twenty-eight or thirty—or something like that—footer. He'd have [help] before the boys started working with him. In the summertime—there was only from June to August [for] salmon fishing. Then maybe a month home, and after that it was crab fishing. [During fishing seasons] they'd come home

every now and then. Maybe stay home a week or less than that. A lot had
to do with the weather. But most of the time they were gone.

Afognak . . . was a village where everybody knew each other, and I had
lived there as a girl so that I already knew the people there. The only time
really [for socializing was] maybe a birthday party, or baby shower, or something like that. I'm part Russian, so I belong to the Russian Orthodox [Church].
We had our holidays, and then the relatives'd visit each other, where we'd
celebrate. It was good. We'd have dinner and then visit. That'd be all.

The first year when I moved to Afognak . . . the children started school.
School was a long ways away, but they had to walk. [It was] a mile and two-
thirds, maybe. Of course we had snow that had started in November and just
stayed there and kept piling, and there were days where the children got
stuck—at school. Then I had to worry about that. I can remember starting
out in a snowstorm, where the snow was just flying [and I was] pregnant. [I]
started to walk [in] snow up to my knees. I'd fall down, pick myself up, you
know. Then I saw a neighbor girl that was going to pick up her family, too—
her brothers and sisters. So she says, "You go on back home, and I'll bring
them all home." And that's just what happened. I don't think I had real
close friends; I was friends with everybody. Most of the people at that time
[in] those days, why, we all had big families. A big family is a close-knit family, like ours were.

*During all the times that her husband was away fishing, Gladys took care of
the children and all aspects of household maintenance by herself.*

There was a time, after I had moved to the village, where [Hans] felt that
I should learn to take care of everything myself. Just made no difference if it
was a man's job. "Don't bother anyone if you [don't] have to. Try and learn
to do it yourself." And I did. I did. I'd shovel snow. [I'd] see that everything
was ready—we didn't run out of water—and [I] took care of my own gasoline
and things for the gasoline lamps. And . . . see that the wood was stacked
up so that we wouldn't run out at night, and things like that. Looking ahead
at all times saved us a lot of sorrow and problems. [We had] no electricity.
[We had] outdoor toilets. The winters are hard, sometimes. I didn't really
feel that I had hardships. I didn't.

There were times where I just had a wood stove, but [in] later years, why,
I had an oil stove. They'd bring . . . fifty gallon . . . drums of oil from the
beach. They'd roll them up and stack them up. Then it was up to me to see
that the oil was pumped into the tank that ran into the stove. That was the
time, sometimes, where I'd stand there trying to open that drum of oil and
. . . I'd have nothing except a wrench to open up their little round things
they call "bungs." You can put your hose in after you open that thing. Well,
sometimes they'd be in so tight that I couldn't open them. Then those little

things that you put your wrenches in so that they catch, they'd wear out. I can remember one time where I got so angry, crying, cold. I hit that thing so hard, and swore, which I didn't do very often. That little bung flew up in the air and I started laughing. I swore, and it flew up in the air! I should have done that sooner, maybe.

Then [I] carried my own water. We had to carry our water from the community well, and it just happened that the community well was not far from where I lived. The . . . older . . . children helped; they had their chores. We had a barrel that we used to fill up with water. We'd go back and forth [for] maybe a day, and we'd fill up that barrel. Then we'd have two buckets that would be standing, and we always had them closed to see that no dust, nothing . . . would get in there.

[We'd take baths] every other day, I guess. We had what we called a wash room. I washed clothes there. [Then] we scoured it clean; I used lye. We [would] heat our own water. [It would take] just an hour. We had a bench, and . . . we had pans, and . . . we just soaped each other and soaped ourselves. [Then we'd] rinse ourselves off. We had a slanting floor so the water would run out into a hole. We could pull out the plug and the water would run out. But we always scoured it, too, to see that nothing was—and my children didn't get sick very often.

[I'd wash] clothes . . . one day a week . . . by hand and a washboard. My daughter was four years old before I had a washing machine. So I raised my whole family, washed their clothes by hand—blankets, everything, sheets— hang them outside. There was no other way to dry them. Sometimes the winter was so bad that they'd freeze. You could just stand them up as plywood or something like that. Anything would . . . freeze dry.

[For clothes for the children] I sewed, and I had no patterns. I'd cut out their old clothes and take a pattern from that. And [I] patched an awful lot— mended their stockings. [Clothing was handed] from the older one on down.

We had no way of preserving food except if . . . we had canned it, or salted it. Everything was either . . . canned or fresh right out of the sea. I canned everything I could think of. Anything that I could get a hold of. Sometimes . . . somebody'd kill a cow in the village and everybody knew about it. Then they'd sell it. So we'd run to buy it to see that we got meat before everybody else bought it. Then it'd be brought home and I'd can it right away, so it wouldn't spoil or anything. Hans would go out duck hunting, and we'd have our ducks. There was a time, too, where later on we'd have elk.

Gladys taught her children the amusements that she had enjoyed in Afognak as a girl.

I can remember how [as a girl] we never were lonely and . . . because we lived in that village, I kind of taught that to my children, too. [I taught them things] like skating, coasting. We had no movies, no television, or anything like that. We had a beach not too far away from home, and that's where the kids used to like to go and play. They'd build little boats; maybe it was part of a barrel or something like that. They'd put a stick on it and a sail [of] paper or a piece of cloth, tie a string on, and they'd just pull it along. As they got older, why, there were chores, and I got them bicycles. They'd ride around town. But they kept away from mischief. Never did I have to worry about that. They played on what we called "The Rockpile" and had a good time. The neighborhood children were free to . . . come to my house, play with my children. There they had coasting [outside], or they had cops and robbers in the house. Our house was small, but I made up my mind, "My children are home. They're safe. Go ahead, bang-bang all you want to."

It just seemed like that was my way of life. I chose it. To me it seemed like that was the way it was supposed to be. I didn't fight it or argue with it, because I felt that was my chosen life. I was contented, and the children were contented.

As in Gladys's youth, Afognak still did not have a high school when her sons completed eighth grade. The older boys were sent to boarding school in Sitka. In 1959 the family moved to Kodiak to facilitate their children's education.

Later on the kids went [away] to school, like to Sitka; they had a good high school there. Of course, they'd be gone nine months of the year. The first year, the oldest one, we never knew [at the time] how bad he felt. He never told us, never wrote to us that he missed us. We wrote letters, and it was a long ways and expensive. The money that . . . [Hans and some of the sons] made fishing, they saved and that was the beginning of them starting school . . . away from home. At first it was awful. There was a time where we had to send three of them to school. Hans and I went into Kodiak with them to see that they got on an airplane. That was a heartbreak. Oh! It was awful to see them go like that . . . knowing that they would be gone for nine months of the year. From there on a few of them went to college, but not all of them.

Then a lot of people began to send their children to Kodiak, to a high school there. We figured, well, rather than to spend all that money to send them to Sitka, and then their tuition and so on like that, why, we could use it. They [could] go to school in Kodiak. So we sold our home in Afognak and moved to Kodiak . . . so that our children [could] go to high school, and yet be home.

Family Fishing

All seven of Gladys's and Hans's sons had the opportunity to become fishermen. Now, three of her sons fish full-time, and two more combine fishing with other

marine occupations. The other two chose flying careers. Neither Gladys nor her
daughter has ever fished commercially, although it has become more common for
women to do so.

[Hans] started training his sons quite early. They all fished with him, except the two that chose flying. [Three] living in Kodiak are salmon fishermen, [and they] crab, and then sometimes one of them would do shrimp fishing. Now their wives stay home. They [own their own boats]. Fifty-footers,
fifty-eight. I think something like that. They purse seine salmon, but then
they crab with the pots.

I have one son in Seattle; in the summertime [he] goes salmon fishing. He
built his own first boat. He went to the Edison—now it's Seattle Community
College [and became] a shipwright. Then he built his own boat. He worked
for different companies . . . that were building [boats], just learning mostly,
at the same time he was going to school. He sold that and built a bigger boat,
forty-footer now. The first one was thirty-two, or something like that. He
comes home after salmon fishing and he builds skiffs. They're fiberglass skiffs,
and he started out [building] . . . thirteen-foots. People who buy them use
them . . . when they're purse seining—when they haul out their net to purse.
Now he's building fifteen-footers; they wanted bigger ones.

Arthur, he went to the Edison School, too. He took up . . . diesel engines, and he got a job right away. He is still in Seattle. He had gone the
last two years and fished salmon or was a cannery boat tender, and they used
one of my older son's boats for a tender. So he was running that. I don't
know what he's going to do now. All this winter he's been working on engines for the canneries, and so on.

I think some of them were eleven years old [when they started fishing].
They didn't do much, but then they began to learn. Now when I look back,
I think their father started training them, talking to them, "Get ahead; try
and get ahead. Help yourself." I can remember where they wanted things
like a motorcycle, which was new at the time, or just a little speed skiff . . .
where they can get out and have some fun. And we'd tell them no. We were
firm. We said, "No. Save your money and buy yourself a big boat that you
can make money on." Now that was our beginning, and that was part of
their training. And to this day, why, they have their own boats. They've
learned trades, which I'm very thankful for. A lot of that goes back to their
father. My boys were lucky enough to have a father who was interested in
them to see that they had a good time. He enjoyed taking them out on a
boat—let them go on a beach. Each one would run to see what he could
find and call to the other ones "I found this!" and "I found that!" They were
quite close together. To this day, they still are the same way.

[My husband never took our daughter fishing]. He didn't believe in that.
He thought that she should learn other things—she's a secretary. [But] we

used to go fishing with him in the spring for home use, and she knew what it was all about. That was more when the children were grown up. I was able to leave home and then sometimes go along with just two or three of us. We had more fun to see all that fish in a net, you know. It was a lot of fun to us. Then, of course, come home [and] it'd be a lot of work, cleaning the salmon, scaling the skins, and then canning. In later years . . . when we moved to Kodiak, we had electricity so that we could have a deep freeze and we'd . . . freeze [salmon].

In the later years, after we had moved to Kodiak, well, then the women began to go out with their husbands. It was the families together on one boat. Maybe the sons and so on . . . and even the girls. Maybe small families would go out and go fishing together . . . so all that money would be in one family. [But] it wasn't too common. Not too many women did that. [Hans] never thought of me ever going out. [But] after the fishing season, Hans and I would go out on the little boat . . . by ourselves. Just go out and sunbathe, and just stay there and maybe pick berries, or just walk around in the woods.

Concerns for the Fishermen

Although Gladys has been a wife and mother of fishermen for many, many years, she continues to express anxiety about their safety. She worried about her older sons when they were in the United States armed forces at the time of the Korean War, but her fears for them as fishermen are just as strong.

My oldest son was a Marine. Then they were shipping them to . . . the Korean War. He went, and then they shipped his clothes home, so I knew they were going to send him. And I figured that's where it would be. But they got as far as Hawaii, and for some reason or other they didn't have to go any farther. So they just maneuvered and he came back to the United States until he finished his term. And then [of] the other boys . . . one was in the army, and two of them in the National Guard. After that it just seemed like they didn't have to go. None of the rest of the boys had to go. I felt like with the older ones, that everyone else is sending their families, and no matter how bad you feel, you have to go through it. You have to be brave. I felt that way, but—I was relieved. I was relieved [that the others didn't go]. I guess there's a time where you felt that it would have been their duty. But I had relatives whose . . . sons didn't come back.

But then sometimes when the boys went out fishing, the bad weather sometimes was just as bad [as military hazards]. They can be out in bad weather across the gulf, and they could turn over. For a moment I have fear. And then I think to myself, "I shouldn't be. I shouldn't be." I think most of us fishermen's wives think that way. We have this little fear for a little while, and then we think, "Why, we have to have faith." I think that's a fisherman's wife's feelings—it's a faith. And *so* many boats are going down and *so* many men are being lost because of bad weather we've had lately. But a fisherman's wife's life is lonely. They're gone. You're alone a lot. But because I had a big family it was all right.

And always when they come home, what a good time. I mean the homecoming. He always went to the children first, because you know how the child [would be], "Daddy's coming!" They'd go running. Well, then, I stood in the background until he was ready for me. He used to say they were the planks and I was the foundation. [In preparation] we cleaned house . . . saw that we had the food to make the special foods they liked, and to see that the children behaved well, and see that everything was ready—do as much as I could. Later on they had radios, so that we could listen in. We'd have a radio at home. Not that you could talk to them, but we'd keep that channel open, so that we could hear just where they are. Then we'd figure, "Well, this is the day they should be coming home." And if it's bad, why, they stayed away; they didn't come home that day.

But there was only one time I can remember [being really frightened], and that was for Christmas. [Hans] was coming home for Christmas. He was crab fishing, and the weather got bad for them, and they couldn't come around into Kodiak. So they went into a bay, and the next day the weather was better so that they started out. It was still kind of bad . . . and a freak wave came and went right over the boat—broke the windows in the [wheel]house, and water gushed in. Hans said he thought they'd never come out. [It] ruined some of the mechanical things they have in there . . . and it destroyed the radio.

Of course, we had the radio open at home. Well . . . we heard Daddy say that they're going back to that bay. And that was it. No more after that. It just happened that my older son was home; he wasn't married yet. Then he and the second son, who was fishing on another boat and . . . was home at the time . . . they got together. In the meantime it had snowed and rained so that the roads were icy. They [the sons] had a long ways to go by truck to go and pick them up and bring them home. They were gone three hours. In the meantime, I didn't hear nothing. They'd call them, but they couldn't talk because it had destroyed the radio. I had cleaned the house. I didn't know what to do. My daughter must have been around ten, eleven years old. She saw me busy [cleaning again]. To her, I wasn't concerned. For that's why I was doing it, because I *was*, and if I'd just stayed and did nothing, I would have been frantic. They came home. I went to the door, and I couldn't open the door, because I had this great big lump that I stood there swallowing. My daughter runs out to her daddy, and she says, "Mama didn't even worry about you. She just cleaned house!" Here I am, standing at the door swallowing the big lump I had in my throat, so happy to see him home. But that's the only accident that I know of that they had. They were just lucky.

One of my sons got caught in a bad winter storm where his rigging and everything got so iced up that they couldn't see, and they ran over a rock. They didn't know where they were going. It just happened that they bounced over. The boat . . . was built down here in Seattle somewhere [and] was built well. It was an iron boat, too, so that it didn't come apart. That's one of those close calls. [Crab fishing] is done in the wintertime where weather is so bad. Cold, freezing, so on like that. They go *way* out to sea to put their pots in today, you know. Before crabs were closer in. [Hans] had the little boat so that he was able to be right in sheltered waters, and yet made enough.

Gladys's sons are skilled fishermen as a result of the encouragement and training that they received in their family.

[People] will say, "Aw, he's just a fisherman," not realizing that fishing is like going to school and taking up another trade. Only you don't go to school for that. You just *experience* it. There's a *lot* more to fishing than a lot of people think. You have to know your tides; you have to know how to set your seine so it doesn't get tangled up in the propeller or cause yourself damage—more sorrow and loss. You have to know about the weather so you don't get to lose the whole seine. Keep your nets mended at all times so that you don't lose that much fish.

[Now], crabs is disappearing, like salmon. But they're building them up now. You know, they're *watching* their seasons. They have a quota and things like that—to get in after that, why, the season is closed. And same with salmon. They're getting so that they have their limits, and then . . . they're talking about hatcheries. The sons that are fishing in Kodiak . . . have been able to get enough to keep them going that they don't have to worry. Of course . . . when they own boats [and] everything that you put in on boats, you get your salmon, [but] a lot of that money has to go back [into the boat]. There's always repairs. And there's the insurance. The insurance is very high. Most fishermen, well. . .Afognak people and Kodiak old-timers, and even the new ones that come from here and go up there to live—have their own boats and then fish year-round.

[My husband] was [cautious]. But he was a steady worker. He worked hard at it. Even if he was not a highliner . . . he always made enough so that we could keep going ahead. I never had to help my fisherman much. He worked hard for what he got. They seemed to look forward to it in the spring, but always happy that it was over with . . . in August month. It's a hard life. They were always happy to come home.

Fish-packing plant *(photo by Sonja O. Solland)*

Marti Castle

Marti Castle was born in Boise, Idaho, in 1947. Her father was a commercial airline pilot. The family also owned and operated a saw-mill. Her mother kept house and helped run the mill.

Early Years and Entry into the Fish Business

I just think of my mother not ever having a job and not ever striking out
on her own and doing things. But actually, when I was little, she and my
dad had a sawmill. And my father was very, very sick for about six months
and she ran the sawmill. She'd go out in the woods and kick the loggers in
the butt and get them going. She hired and fired people. It's a side of my
mother I can't relate to at all. I was only, maybe, three years [old], maybe
four, so you don't have much of a memory at that age. That's the only job
[she had, except] I think she had some job during the war—making maps or
something.

In her own way [my mother] is very independent. She's got a lot of stam-
ina. There were five kids in my family, and we would go to stay at a lake in
northern Idaho in the summer where we had a cabin. It was on the far side
of the lake. We didn't have any electricity or running water and [we had] a
wood cookstove. She would be there with five small kids, and my dad was
not there that often. There's a lot of bears around there; I never thought
anything about it. [But] other women I met say, "I'd never do that. You can
forget that, with five little kids and bears out there."

We were always taking off in our little dinghy that was half sinking; she
wouldn't care. My friend and I would take our boat, which was an eighteen-
foot boat, and . . . go on little trips. That was [when I was] twelve years
old—not even that old—ten. We'd go off for the whole day with the boat.
We'd beach the boat and tie it up and go off doing something. One of my
friends has got two kids who are thirteen, and she doesn't even want them
to stay home by themselves. She'll leave them for a little while, but they
have wood heat and she worries too much about it. I think, "Geez, here was
my mother letting me take off with the boat and . . . Priest Lake is at least
as big as Lake Washington." I don't know if I'd let a ten-year-old do that.
But my mother has always been that way. So I really have to give her a lot
of credit for that.

*Marti's father was transferred from Boise, Idaho, to Spokane, Washington. This
presented a dilemma: "he could either quit flying and keep the sawmill or sell the
sawmill." He sold the mill, and the family moved to Spokane, where Marti and
her family lived until she was fourteen. The family later moved to the Puget Sound
area, where Marti graduated from high school. Marti got into the fishing business
by chance. She was living on a farm in Enumclaw, Washington, near Seattle—
raising goats, weaving, and, as she says, "being very poor."*

It was absolute chance [that I got into the fish business]. In fact, I was on
my way to moving to eastern Washington, and my sister talked me into going
to Friday Harbor. I had vehicles to sell, and animals to sell, and a lot of loose

ends to take care of. Everything went wrong. I had the hardest time. It took me about four months to finally liquidate everything. It should have taken me about four days. My sister had a friend who worked in a salmon cannery in Friday Harbor. She needed a job for the summer when she was going to school. So she talked me into going up there with her, and we both started working there. I ended up working in the office, which I didn't want to do. But I did [it] because that was the deal they gave us. If I worked in the office, [my sister] could have a job. She was working in the processing. I ended up staying, and eventually [I] became the manager of the plant.

It's called the J. J. Theodore Company. Over the years of working there, I saw their production go to a peak and then start dwindling off as they lost boats and as the fishing time in Puget Sound was cut down. They're still there, I think running pretty minimally. I think they now do some custom work for a Canadian company. It's real hard in Puget Sound, because they fish so little that a plant can hardly keep going. It was a cannery, and all the freezing that was done on our fish, we had to send it away.

At first Marti's parents were not pleased with her choice of occupation.

I think my parents had a little bit of a hard time with it at first. They didn't feel that it was a fit place for their daughter to be. But they're very proud of me now. I'm sure that what it was, was that they had their own ideas about what I should be doing. It didn't take them long, but it took them a little bit to get used to it. Until they knew that that's what I'd decided I wanted to do. Then they were very supportive. Now they think this is great. My parents have always been—that's something that was sort of drummed into us—being very achievement oriented, very competitive. I think all the kids in my family are that way. They're all achievers and pushers. So I guess that drive was probably always ingrained in me, although I never really went out and did it. They came up to Friday Harbor to visit once. They walked into the plant, and I was driving the forklift around. Right off the bat, that's not the idea they had in mind for their kid. But, as a manager you can't just sit around and watch people. You've got to jump in there and work sometimes, too, doing whatever—driving the forklift, or sliming fish, or driving a truck.

After working in Friday Harbor for six or seven years, Marti found managing the plant less of a challenge and decided to make a change.

Friday Harbor finally got to the point where it was not a challenge anymore, because the fishing in Puget Sound had gotten worse and worse, and the business wasn't growing. A couple of weeks before I was going to leave Friday Harbor [I called Sea-Pro] just to get some business straightened out. [I mentioned that] I want[ed] to get it done . . . because I was leaving. [They asked] "Where are you going? On vacation?" I said, "No, I've quit this job." "Where are you going? What are you going to do?" "Well, I don't know, I haven't figured that out yet." "Well, will you come and talk to us about this?" They were just starving for some good people.

Managing an Alaska Processing Plant

Marti then went to work for Sea-Pro, a freezing and cold-storage operation with plants in Seattle, Washington, and Anchorage, Alaska. At Sea-Pro she found the challenge that had been lacking at Friday Harbor.

[Sea-Pro] was a much bigger operation. The freezing part was something I knew nothing about. I always wanted to work in Alaska. I always wanted to work with a freezing process; it was something that I'd never done. So it was great. [And] this company is always trying to expand. At least if it's not doing something different, it's always trying to push more product through. So, that keeps you going. When I was in Friday Harbor, the biggest year I can remember was [when] we did about a million pounds of fish, and that was on the round-weight basis. Up there we do probably eight million pounds. It's a big difference. [And we do that amount] in probably about the same time period. Even [doing] eight million pounds, it all comes so fast that there's a lot of time, early and late in the time we're up there, that we're not that busy. But when we are, you're real busy. [Also] in Friday Harbor you're shut off from other plants at other places. Whereas working as a custom processor, you're more aware of all other plants, whether they process their own fish or not. See . . . how I knew of Sea-Pro was through [Friday Harbor] because that's where we sent our fish sometimes to be frozen; and that's how they knew of me, too.

Marti and her partner are responsible for all aspects of the processing operation at Sea-Pro in Anchorage. They constantly make decisions about the operation of the plant, the intake of fish, and the performance of employees.

[Now] I'm one of two managers. [My partner Tom and I] go up [to Anchorage] the first of April and set up and run the whole operation [and] come home in September. A lot of people would say, "Well, I don't see . . . [what] they send two of you up there for," because sometimes it's not enough for two of us. But, for a lot of the time one person would lose their mind trying to do it. One person couldn't do a good job. So, it's better, and it's been very interesting working directly, equally, with someone else. Neither of you has authority over the other one. It's worked out very well. [My job includes] scheduling in business, taking care of all the billing and receiving of the money, and making sure everything is running right, [and] hiring.

[Sea-Pro] is a custom processor. [Fish belonging to other companies] is trucked to our plant, and we butcher the fish, freeze it, glaze it, and ship it to Seattle for storage, because . . . we have very little storage there [in Anchorage]. So, what we do is try to find the business when we're real slow. We contact companies we know that are out buying fish, to see . . . what they've got going for the day. We have a certain number of customers that are constantly bringing fish in. We'll keep coordinating with them, so that we don't overload ourselves, and so that if they have fish to bring in, and if they bring it in right now, it can get done. If they bring it in the morning, it's going to have to sit for quite a while.

We don't buy any [fish]. We just process. [The fish are flown or trucked in]. There's a lot companies that have no processing plants of their own. Unless you're running a real high volume and running year-round. I don't think it's really worthwhile to build your own plant, because you put all that money into it to use it for a very short time. So, if you can find somebody to process it, and you have the guarantee that they will process all of your fish, then it's more beneficial to do it that way. Unless you want it all canned; in that case, usually you can find someone to can it for you. Companies like Seward Fisheries do custom canning. From us it goes on . . . ships in refrigerated vans, and it's brought down to Seattle. It goes into our plant's cold storage here, which is a big plant called Sea-Pro Seafreeze. It's on the Duwamish Waterway. It's stored, and then [the owners of the fish do] whatever they want to do with it from there.

During our busiest time, about one hundred and fifty [people work in the Anchorage plant. There are] probably more men [than women among the employees]. It's probably sixty-forty. In our plant we had . . . three foremen and two supervisors. Two of the foremen were women—and one, she was incredible. She is now managing the plant for this guy that we've leased it to for the winter, who's processing shrimp. But I think . . . [it's] true usually [that] it's women working on the line. These particular . . . [foremen] take care of a certain section of the processing.

We have two different buildings. They're kind of in a little cul de sac, so you have to go across the road, across a little tiny street [to go to the other building]. For our processing, it's divided into the butchering and cleaning of the fish, and then putting them in the freezer. Then when they come out of the freezer, they go to the other building, where they're glazed and packed in boxes and then put in the vans. Sally was the foreman over the glazing area and took care of [it]. That was her thing. If she needed new employees, she was free to hire them if she wanted to.

But, then there was another woman whom I really supported strongly as a foreman, and she just ended up not doing a good job. [Tom and I] work our hours. Tom comes in about 6:30 [A.M.] until probably 6:00 or 7:00 in the evening, and I come in at noon and stay until midnight or sometimes later, but thereabouts. After I leave, I don't know what goes on. [Marlene] would go off in the night, leave her crew, and go out in the car and get drunk or get stoned. The day crew would come in the morning, and [she] would be just totally screwed up. Everything would be a mess. I was the one that supported her, and I was the one that finally got totally ticked off at her. And I was the one that fired her. That's something I've never done. I always feared I would have such a hard time firing somebody or doing that. I thought it would kill me, because she was sort of a friend. All day long I was sick to my stomach, thinking about when she came in that night, that this is what I was going to do. It turned out not to be quite as hard as I thought. I guess I could do it again. I didn't fire her. I just told her . . . if she wanted to continue to work on the line, then she could, but she wasn't going to be a foreman. But one thing that was always drummed in my head when I was in Friday Harbor by the owner of that plant was, "Don't be friends with people that work for you." There's certain people that will never take advantage of you. But, by and large, it's better to keep a bit of distance so that when you have to do something like that, you're not doing it to your best buddy. Because that would probably [be] really, really hard.

Then [there's] just the workers generally. Tom and I would go through all the applications and just call in people. And then . . . they'd just get stuck somewhere. Usually no one is all that capable [or has such] great skills that you're going to say, "Aha. Put this person here." Unless they're a good forklift driver, or . . . very good with a knife, or . . . good with quality control—those kinds of spots.

Most of them that work [at Sea-Pro in Anchorage] are under twenty-five, I'd say. There are a few that are older, but hardly any. And just all kinds of people. I mean we've had people from every nationality you could think of. Asians aren't the greatest [number]. There've been some Mexicans, some Afghans, some guys from Vietnam, Korea—I don't think there've been any Japanese—Filipinos, a couple of people from Italy. Just from everywhere. England [too]. Anchorage is a real melting pot. Half of them can hardly speak English. It's real hard for them to understand. A great mixture of people. I could never ask for a better crew. Just real good workers. Most of them [the crew] are residents of Anchorage. Everybody's got a brother or a sister or a friend—or their brother's friends—who want to come up there and work. If they do, they've got to find their own place to live. That's up to them. Housing [in Anchorage] is rather expensive, so for a lot of people it's not really worthwhile. Usually places that do have a bunkhouse, they charge you a certain amount for room-and-board a day. If you're not working, it gets hard. You're not making any money to pay for your living.

We went through a really slow-period last year. It was well over a month. But there was just hardly any work at all. I expected that we would lose our entire crew, which we didn't. We lost the people that had to go back to

school, but most of them just stuck around, which was really nice, but . . .
they weren't making any money. But, on the other hand, they kept thinking, "Well, maybe, maybe there'll be some work." I have a friend who, as a
kid—his father was Filipino and he went up—would go to Alaska with some
of the old Filipino guys and work. I don't know how young he was when he
started out working up there. He said that even when he got out of high
school . . . if they wouldn't give him a leave of absence, he'd quit his job
and go up there in the summers. He doesn't do that anymore, but I think
the lure is still there. He still thinks about it.

*In addition to fluctuations in the amount of fish available for processing, and
other unforeseeable challenges, dealing with desperate customers and "shifty char-
acters" is part of the job of a manager in the fish processing business.*

We have quality control that check every load and take temperatures as
they come in, and that's some kind of protection for us. If they found really,
really poor quality fish, immediately we would get a hold of the customer and
do something about it. But that very rarely happens. Or, we have the option
of not accepting fish.

Once we had a guy who drove in—he drove in, in the middle of the night,
with a big trailer full of fish. It was about 4 A.M. and so the foreman figured,
well, he'd just wait. [He knew that] I'd be there at 6:30 or 7:00, so he didn't
call me and wake me up. But the guy brought them. He wanted them processed. Well, they were all rotten. When I got there, I just told him no. We
had too much other stuff to do, without messing with fish that were already
rotten. And, oh, he begged and pleaded. And then he was out there all day
long with his van, having people come and look at his fish. He brought them
from Kodiak into Homer, and by the time they got to Homer, they weren't
doing too well. He tried to sell them there, and nobody would buy them. So,
he unloaded them and trucked them to Anchorage, and finally he did sell
them for peanuts. I think a woman bought them for a penny a pound or
something. Well . . . I don't know what kind of plant she did. It was something in her garage. She took them and went through them and pulled out
the good ones and cleaned those and iced them. Then she brought them
back, and we froze those. She was going to have them canned down here in
Seattle, and I don't know, they're still sitting in cold storage. They've been
there for two years.

[One of our customers tried to bribe me once.] I've talked about it with
my boss, because . . . nobody quite knows how to deal with it. I don't know
that it's never happened with anyone in our company. I was talking with my
partner in Alaska about it. He said he had never been bribed, but he knew
a lot of people that had been and . . . [that] somebody did try to bribe him
once. I couldn't do it. You start doing that and you're sunk. You've got to

keep it up then. Once you've done it, they've got something on you. I said [to the customer trying to bribe me], "We're here to perform a service for our customers," I said, "not to take bribes."

Marti and her partner run the Anchorage plant from April until September. They then return to Seattle and work in the company's Seattle plant for the rest of the year.

In the winter, both of us that go to Alaska kind of just do whatever needs to be done [in the Seattle plant], wherever there's a spot that they can put us. Last winter, a fellow who managed the frozen production in a plant down here quit, so I took that over until they could hire somebody new. This winter, we leased a plant at Pier 89—New England's old plant—and both Tom and I worked there. I worked with processing crab, and he worked on the cold storage. [I worked] on the production end, over the people on the lines. What they do in Alaska—after the crab is brought in, they butcher it. It's pretty much just breaking the crab in two. The [body] shell breaks off so you've got it in two pieces. That's frozen and packed in varying-size boxes. Usually they go from forty to eighty or a hundred pounds. [The product is then shipped to Seattle]. When they're ready to sell it . . . those halves are cut into sections, and [then] they're cut up . . . into individual legs. Then it's packed that way. Sometimes they size it, and sometimes not. But it's packed with a ratio of legs to claws. Well, what we've got right now [is] king crab. And then opilio, we've been doing. Or sometimes they split the legs in half. Sometimes they extract the meat. All kinds of things.

In the winter, usually [the processors] box up all the fish that's been caught in the summer, and . . . it's sold . . . during the fall and winter and spring. So it's boxed for shipments and packed. They pack it in individual polybags. Either 50-pound, 100-pound, or—like big kings are usually packed in 200-pound boxes. They're boxed up. So, what you're doing is just emptying out your cold storage, pretty much.

[This year, while I've been on vacation], I think they [Sea-Pro] were doing a big project for Peter Pan on portion control. I think they were taking frozen fish and thawing it out, and then filleting it and cutting it into a portion size—however big a size they wanted, probably four or eight ounces—and then [packing it] in a shrink-wrap.

The change Marti must make every year from manager to worker is difficult for her.

[My partner and I] are management people, and that's our function in Alaska. But coming back down here, the plant here already has its own management people; so we just kind of do whatever needs to be done. It's real hard going from where you're directly responsible for everything that goes on, to letting someone else be totally responsible. In Alaska, it's just taking a whole plant and making it run, making everything go, making sure everything gets done—that all the customers are happy, and that, from the production standpoint and the administrative standpoint [it's going well]. It's a big job. It's an interesting switch, when you're not in charge. For me, it takes some getting used to, and then it's okay. It's nice to let somebody else be responsible for everything and to just walk out the door and say—"It's their problem." But it was real hard at first not to butt into their business and say "Why don't you do this?" and "Why don't you do that?" and "Why don't you do it this way?" [I was] seeing other ways to do it. And [then I was] just feeling—this is the person's job, and I'm just here helping out; so let them do it their way.

Over the years, Marti's sisters and one of her brothers have worked in the fish business periodically. One of Marti's sisters, Marti, and her partner have all experienced physical ailments common to fisheries workers.

Both of my sisters have worked in fish processing plants. One of my brothers worked at the Friday Harbor [plant] for a couple of months, and the other one never has. My brother who worked there for a little while . . . he wanted to be an air force pilot. So he was just waiting to get into a flight program. My other brother's been a merchant marine for a long time. He's never really had any interest in it. He's on boats all the time. I think my sister [that] I went to Friday Harbor with originally—she had a good time, but she was going to school to be an occupational therapist. That was what she really enjoyed. So she just did it kind of for the heck of it. My other sister likes it, but she likes art work better. [She] has worked up in Alaska. She came up and worked for me last summer. She's worked in Petersburg, and then last summer she also went to Cordova. There were about six people that went to Cordova. One by one they all came back. They were wailing and moaning about how horrible it was. She wanted me to call her right away. I was prepared for this sob story. She was as happy as could be. She said, "I don't know what their problem is." She said, "Yeah, we're working sixteen, seventeen hours a day, but they went there with that knowledge." But none of them could make it.

I don't think she can do it anymore. She got tendonitis. It's an ailment of some kind that gets your tendons, and usually they start contracting or they get very swollen up. On her hands, her middle finger is the worst. She can hold her hand straight up, and this finger just starts twitching. It's real weird. It just goes like this, until all by itself it just curls up like that. On one hand it does that, and on the other hand it will go a little ways but doesn't look nearly so bad. It's [from] the cold and . . . she worked with a knife a lot.

She . . . just constantly was hanging on to that knife. It's just a constant strain on the tendons in one part of your body. A lot of people get it. They get big splints on their arms. That's one way you can get rid of it. It comes from overuse of a particular muscle.

My partner up there got it once. They were grading a bunch of frozen fish down here, and he was just helping out, putting the fish on the table. He was picking them up out of a box and throwing them on a belt, like this, all day. Well, he got it in his arm, and he had to have a big splint kind of thing on his arm until it went away.

I've [had problems with] my elbows before, from hanging onto fish, or from I don't know what. I was slitting bellies with a knife. It wasn't even for that long. It was probably for six or seven hours. They're better now, but for several months I couldn't carry anything. If I was carrying a bag and my arm was pulling down, my elbows would start hurting. And they still do a little bit, once in a while, but not too much. But it's just that constant strain, and the cold . . . because it's always cold inside of plants. There's cement. And you're standing on cement. You can't turn the heaters on, because that brings up the temperatures of the fish and they spoil faster.

Reflections on Women in the Fish Business

When Marti first started working in Alaska, she had to cope with both her own reactions to being a woman working in a male-dominated industry and to the customers' reactions to her.

Women . . . in the fish business, they're not that common, except as traditional secretaries or bookkeepers. In Alaska [even] most of the bookkeepers in canneries are men. They're almost always men. A lot of your office staff are men. Women will be floor ladies or kind of a floor foreman. That's a pretty common spot for a woman. I think in our plant, just having one of the managers be a woman, there's a lot of women . . . they feel easier about going ahead and doing stuff; [like] learning to drive a forklift. They'll go ahead and do it. Whereas I know down here at our plant in Seattle, forget it. You never see a woman on a forklift. The boys won't let them on. Actually . . . there's forklift drivers and that's all they do. They don't like anybody driving their forklift. But up there, we've got a looser operation in those kinds of senses. So I think women are able to do more things. They know that nobody is going to say anything, and they're not going to be told that they can't do that.

I think a lot of people automatically assume that a woman doesn't know as much, and therefore they'd rather deal with a man. This I definitely see in Alaska, having a male partner up there. A lot of times people would much rather go talk to Tom. [That happened] until . . . it became apparent that

I did know what I was doing and that I could take care of their problems too. Over the years I've lost that drive that says, right off the bat, "Look, I can take care of your problem, you know. I know as much as he does." Now I don't care. I figure, if they don't want to listen and they don't want to talk to me, I don't care. That's their prerogative. In the end, they'll find out that I know as much. If they don't choose to talk to me right off the bat, I don't take it personally anymore. But it is a battle. Men will always—well, anybody would rather talk to a man. They just assume he knows more. I could just say in our particular case, Tom has worked in the fish business his whole life. His father has always had businesses. So, in a lot of respects, he knows a lot of things that I don't. But they're not things that are directly related particularly to what we're doing. He's just been around it longer. I think, too, that men like to just sit around and bullshit with another man.

Over the two years I've been in Alaska, I've seen a change in [the process of] dealing with the people we deal with. They'll deal with me now, too. I get along with them just as well. It just takes time. There's always a few people that I won't deal with. Period. That's it. And Tom will. We have a lot of unspoken agreements. He knows who I don't want to deal with, and he will do it. I know who he doesn't want to deal with, and I'll take care of those people. Or I know there's certain things that *he* does *not* want to deal with; just things that have to be done. I'll do those. He knows the same of me. There's just some very subtle understandings.

There have been times when somebody's come in [who] doesn't feel like a woman should do certain things. [Tom] will just kind of stand back, and he gets this grin. He just kind of chuckles, because he knows what that person is asking for. He'll just sit back and watch . . . me just "light into" somebody. He doesn't say anything. I think he gets a kick out of it, because I don't think he's ever worked directly with a woman before. So that's a new thing for him. I think, too, sometimes [in] working with a man it becomes difficult because you spend so much time together [and] then that causes resentment from his wife—which has never happened [with Tom's wife]. His wife is really very, very nice, and that doesn't seem to have ever created problems for her. His wife is totally different than I am. She's real traditional—stays at home. They have three kids, and they're all pretty young, so that's a bundle to take care of. I've always tried to be considerate of her.

I've had to deal with that before when I worked in Friday Harbor. The manager before me, his wife really resented me. I'm sure for her it was [with] good reason, because Leonard would stop by our house all the time. He lived not too far from where I did. He'd be driving by, and if the lights were on, he'd stop. He was my age but had never gone through the same kinds of things that I had. I don't know how old they were when they got married, but I think he'd led a very traditional life. The whole late sixties, early seventies was something he'd totally missed out on. He had no consciousness of the music of the times or any of those kinds of things. He was so curious about what other people his age did rather than stay home with their wife and kid. Anyway, he would stop by and just chat. Sometimes he would borrow records and stuff. His wife didn't like that at all. There was just a lot of resentment there. I think, in some ways, it caused problems working with him. But from that experience, I never really wanted that to happen again. And I'm glad it hasn't.

[As for employees], it's real interesting. There's a lot of resentment from men working under a woman, often. I've seen that a lot. I've always tried not to be insulting to the guy, but sometimes they don't want to be told anything by a woman. And other people, they could care less. I think that a lot of workers try to take advantage of a woman boss more, too. [Then too, sometimes they can be flirtatious. Once] there were two—they were really young . . . only sixteen or seventeen—and their father was one of our supervisors. They were always real flirts, and they were both really cuties. But they'd always come up—they were just so silly—and try to flirt. There was one kid, he would bring me flowers. They were just real funny.

Sometimes you expect in a job more sexual harassment from people. I've never had that happen. But I don't let it happen, either. If there's any indication, I always just keep that person at a distance, because I don't care to deal with that. Not a distance where you're not friendly, or you can't be friends with people, but enough of a distance that you don't let that sexual edge sort of slide in.

Retrospectives

Marti enjoys the fish business, although she wishes she had more time for weaving. Over the years, her confidence in herself as a person has grown through her work in the business.

[I] just fell in love with doing this fish-business stuff. I love it. I don't know why I love it so much. I think partly . . . it's always a challenge. [What I like is] that it's something that changes day to day. It's the same in one sense, but it's always changing. You can never be sure of what's going to happen tomorrow. Tomorrow morning you might walk in and your plant is just plugged with fish. How do you take care of them all without them spoiling? Or you take the risk that somebody's kind of slid in some fish that aren't very good quality, and [then] blames you for their spoilage.

I never had any aspirations to be a manager of anything. I don't think, in the past, I ever recognized having that drive that it takes to get there. I always tended to have low self-esteem. But then I see that there's not that many other people that do [that great a job]. [This guy] came to work for Sea-Pro. I always had this guy set up as an idol, in a sense. He's done all these things, and he really has a lot of knowledge, and he must be real hot stuff. Well, I worked with him this winter and I was very disappointed. Here's this person who, I thought . . . just knew so much and was so skilled at what he was doing. He did some real dumb things, I thought. So [it was a matter of observing] some other people who I had felt would do a far superior job than I would, and seeing that they don't.

I think ideally I would like to work just in Alaska and not work down here [Seattle] in the winter, because I don't really enjoy it. If I was really needed, that's one thing. But I feel like they've got to find a job for me to do, and they could pay somebody a lot less money than they pay me to do the same job. So, it's sort of stupid, when I don't really want to work in the winter. I have other things I'd rather do [like my weaving]. But that was the deal that I was hired under.

I think about [what the attraction to the fish business is] for myself and . . . it's nothing I can pinpoint. It's just something . . . that gets you, and you have a real hard time giving it up. I mean, it's dirty, and it's stinky sometimes, and it's . . . real unpleasant. You work extremely long hours. I can think of days where I've not slept for two or three days. I'll go home for a few hours, but . . . I'm too keyed up and I can't sleep, so I just go back to work. But what is it, that you'd push yourself to such lengths? Certainly the money is not it. I don't know.

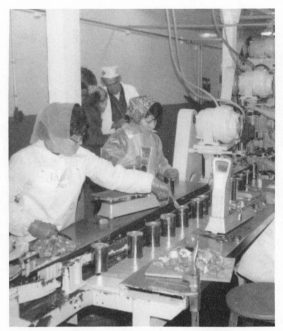

Fish-packing plant *(photo by Sonja O. Solland)*

Christina Jefferson

Christina Jefferson was born in Bangor, Maine, in 1953. She is the eldest of four children. Christina's father was in the Air Force, and the family lived in many different places including, for three years, Ankara, Turkey. Christina's mother met the challenge of the Air Force life-style. Christina's father retired in 1962, and the family settled in her mother's hometown in Delaware. Her father continued his flying career with the Civil Service.

Growing Up

I was really lucky because my mom was a pretty cosmopolitan person. She's not one of those women to stay in the United States and say—"Oh you go over there [to Turkey], but I'm not going to go." It's drag all three babies over there and put up with whatever life had to hand to her. I have a lot of respect for my mom, because she was really behind the man all the way, and he was one of those guys who flew his maximum every month—the professional pilot. He was eleven years older than she was; so it had to be love, I guess. I like to think it was. [Maybe] everybody likes to think that. She went through a lot over there [in Turkey], but she always got along. She was pretty tough in a lot of ways. [She was a nurse.] She nursed a year before she married my dad. Then I came along a year after that. She never went back to nursing; although, in a sense, my mother's always been a nurse and always will be a nurse. She'd go down into the slums of Ankara, and . . . use her privileges and PX's and stuff, to take food out to give it away. She could have been put in jail . . . because, in Turkey, they don't like you to do things like that. They ordered us to burn clothes . . . that you wanted to cast off . . . because you weren't supposed to give it to the population. She's just a really good-hearted person. She'd give you the shirt off her back.

My dad, he was a pretty interesting person. He wanted to go to college, but his parents [said], "Why should we send you to college? Go out and work for a living first." So my dad joined the Air Force. He got elected to go to officer candidate school, went to officer candidate school, and did his thing. [He] flew hospital ships in the Korean War, flew all 118 missions of the Berlin Airlift—just really hot into these things. His hobbies? He was a really great photographer. In fact, we have tons of his slides at home, and if we could put them together they'd be some great anthropological study. [There's] no country he ever went to where he didn't do an in-depth study, from the bottom up, of any kind of people. Like in Turkey, he'd spend . . . all weekends—he'd drag us out, all three of us kids. We'd get on the scooter, and we'd go out into the hinterlands, and he'd make all kinds of neat studies about people threshing wheat in the ancient style.

[My parents] encouraged us to look beyond where we were and to look towards the higher things in life—reading good books, listening to good music, the whole thing. So much so that . . . I never heard a whole Beatles album until this summer. I never knew what country music was until . . . I went to Dutch Harbor. [My father] was really pretty Victorian in some of his modes and manners. Kind of away from us and yet with us, because he was older, because it was hard for him to relate to children. He was quite an influence, in a lot of good and bad ways, like all parents are.

My dad was Catholic; we're Catholic, but we're not good Catholics. My mom . . . was Methodist to begin with, and then Presbyterian . . . and then she converted to Catholic when she married my dad. She couldn't marry my dad unless she converted . . . because my [paternal] grandmother couldn't see my dad being married to a non-Catholic, and the child was going to be born in sin. They've been married for a year . . . and my mom gets pregnant, and my grandmother [says], "She'll be born in sin. You've got to get married again." So they get married again by the priest. But we were never brought up in a strict Catholic faith. We were baptized Catholic, and that's it; we weren't even confirmed.

[My father] was of Lithuanian descent. His parents came over—they stood up all the way over to this country. [They] came to this country with nothing and made fifty cents a day in sweatshops, and shoe factories, and things like that. They built up their little nest egg in this world. They're pretty determined people, which to me is—I don't know, sometimes I think I could never be worthy of this kind of strength they had; it was just pretty tremendous. I never realized how important my grandparents were to me.

[For] my old country grandparents . . . [I felt] the same way. My [maternal] grandfather, he ran a farm. He put up with dairy cattle [and] did a thing with horses. He had seven children in the middle of the Depression, and he still came out on top of things. So both sides are pretty determined people.

In high school I decided I'd never make it into college because of my math grades, especially in geometry, because I flunked. And I figured, "Oh my God. I'll never get this; so I've got to go out and do other things." So I ran my own campaign for student body president, and I got elected. Wow! I beat all the guys! All the big football-hero team people! Ran my own campaign. Had a good time. I was always into theater too . . . like student directing and acting. [Also] ended up I was vice-president of [the] senior class. So I figured all these neat extracurricular activities [would] give evidence of my great character; that was the whole point. [Then] I was elected to the Honor Society. I felt they just elected me because I was student council president . . . so I told them they could take it and shove it. I didn't like patronization.

I really didn't know what I wanted to do when I left high school. I cried in my advisor's office, because I really didn't want to go to college at all. But then I didn't have any other thing or place to go. You're educated through academic courses, and therefore your next logical step is college, and wham! I decided on the University of Alaska. I wanted to get as far away from home as possible. I hate my hometown with a passion; the only reason I go back there is because I love my family.

When I told my father that, he just blew up. "Oh, it's just not practical. You can go to the University of Delaware, and you can get a B.S. in nursing. That's what you can do. Don't you want to be a nurse like your mom?" [I said], "No, I don't want to be a nurse." He goes, "Well, how can you say that?" "I can say that because . . . I candy-striped in the hospital from the time I was twelve years old." Because I'm my mother's daughter, and because everybody in the hospital knows my mother, I got to do all kinds of nifty things. I . . . didn't want to be a nurse. I felt like I couldn't, because if I gave and gave and gave, there'd be nothing left. And I'm going, "Oh, God. I'm being selfish." But that's what I had to be. I decided, "Okay. Well then, why [not] go to school and major in something like sociology or anthropology?" I could have gotten off sitting in some corner someplace writing notes about people. [My father] didn't want me to go; he was really upset with me. You know, the old father-daughter thing. I didn't talk to him for three months before I left. But he still let me go. I couldn't have gone without him; it was his money. I kept going, "Well, I'm going to get a job, and I'm going to pay you back." So he put me on a plane to Fairbanks, and that was the last time I ever saw him because two years later he died. [It] blew me away.

First Seasons in the Fish Business

Christina was first introduced to fish processing in 1972 at the end of her first year in college at the University of Alaska, Fairbanks. What began as a summer job to make money became much more to Christina; she did not return to college until winter quarter. She went to the Aleutian Islands, and in her first six months in fisheries Christina processed fish at Dutch Harbor, Seldovia, and Adak.

I got hung up the first summer out of school working in the fish business. I heard it was unskilled labor and you could make good money at it. A girl friend of mine said, "Hey, we'll go make some money in a cannery." She wanted to go to Seldovia. We had gone down and applied, and we got sent applications for Dutch Harbor. Well, her father didn't want her to go [to Dutch Harbor] . . . because it was a whole different atmosphere than Seldovia, which is a small boat fishery. And I went, "Well, I want to go. I've got to go someplace I've never been before." So I flew to Dutch Harbor.

Christina's first impressions of Dutch Harbor were not encouraging.

I stepped off the plane in Dutch Harbor in May . . . 1972. A small plane—it was a 410, seated eight people. It was snowing and raining and grey and rotten, and I looked around me and—Dutch Harbor—there was a ruin there. I mean it was nothing but old ruined military installations. And I'm going, "Oh my God. What did I do to myself?" Then my boss, who is a really good friend of mine now . . . probably he was about forty then, and [with] just piercing black eyes. Meanest look I've ever seen. He brings this beaten-up pickup truck to pick us up, and there were—I think there were seven girls and one guy. Anyway, he's hunched over the steering wheel and we're crammed in the front seat and we're going down this dirt road and all these buildings are collapsed around. He looks at me and he goes . . . "Dutch Harbor is exactly what you make of it." That's exactly his words. And by God, he was right. He was never more right. At first . . . oh man, I thought I'd come to hell. Until you started looking around you and then you realized that you couldn't be in a more beautiful place. There are real eagles, and there's real seals, and there's real whales, and there's real snow and real mountains and real Indians and the whole thing. The Aleut people—of course, you probably shouldn't say Aleuts are Indians—they are their own special race. Dutch Harbor is on one side of Amaknak Island, where . . . most of the boats are tied up. There's a bay, and the town of Unalaska is on the other side, and there's no way to get back and forth. So you are stuck on this island.

When Christina first worked in Dutch Harbor, the ratio of men to women was about fifty to one. All the women workers were subject to a lot of sometimes unwelcome male attention.

In the next couple of days we went to work for Alaska Foods . . . [which is on] this 425-foot Liberty ship; [it] has rust 20-feet thick on it. You step on board this boat and . . . you just went, "Oh no!" You walk in the galley— twenty zillion guys all over you, because you are the seventh woman to ever set foot on the boat and they're all "AAAH!!" It was bad. They'd all been there for two months. So [that] was a really strange [experience for me]. We were working night shift, and we were processing shrimp, and we were staying on this greenbelt picking all this shit out of little shrimps. Oh my God! I mean all this assembly-line work, [it was] freaking me out. I didn't know what was going on. But you just work hard. I mean you stand up X amount of hours and pick as much shrimp as you can pick. So I just got really into the whole thing, and had a really good time, and learned all about production.

Alaska Foods has a really good name in the business, and I was very fortunate. The bosses that I had were all married. They all had their wives there. The bosses went home to their thing at night, and . . . they never put anything upon you. [But] Dutch Harbor was heavy for a nineteen-year-old. I was nineteen when I first went to Dutch, and it was heavy [for me]. I can see why women go the way they go in Dutch sometimes. Because if you don't have any kind of foundation, or you don't have . . . a good self-image, you can't make it in Dutch Harbor.

One day I was walking down the road and this one guy on this fishing boat just goes, "And who are you?" He was pretty neat. But I wouldn't give him the time of day because my upbringing was pretty . . . puritanical in a way. Where I come from you never go in a bar unaccompanied by a male, and you never go in a bar if you are under twenty-one years old. No way. You don't smoke. You don't drink. You come home every night. You just do normal things. Well, he came up and talked to me, and I just walked right on by. Later on—it took me three weeks or four weeks to get to the Elbow Room, which is the only bar across the bay. I didn't want to go, because it's a big thing [with me]. I didn't drink anyway. So I met him in the bar. He was my first fisherman trip. I was in love with him for two years. That began my big love affair with Dutch Harbor.

After three months in Dutch Harbor, Christina decided to try to make more money processing fish elsewhere. She went to a cannery in Seldovia, and made plans to try crab processing in Adak. She was on a contract at Dutch Harbor, which guaranteed her plane fare and a bonus if she stayed there for six months.

In August I got really upset at Dutch Harbor, because it was no money, and I couldn't see myself going back to school and getting all my money together. So I split Dutch Harbor and went to Seldovia, which is on the Kenai Peninsula. I worked for a month in a cannery with a friend of mine. Then their king crab season wasn't anything like the one that was coming up in Dutch Harbor, so I went back to Dutch Harbor. Well, I figured I'd forfeited . . . my six-month contract [with Alaska Foods], which [would] mean I get my plane fare and a bonus [at the] end of six months. I figured I'd forfeited that already, and that's why I made plans to go to Adak. But [my boss] wanted me to stay in Dutch Harbor because he had two big loads of crab coming in. I said, "Well, I assume my contract's null and void." And he goes, "No, I just gave you a leave of absence. Don't worry about it." So, that's fine. But then came the time after the two crab loads were fully processed, and there was nothing else in Dutch Harbor, and I was going to Adak regardless. The morning I'm ready to leave [he says], "I can't give you your plane fare or your bonus." I couldn't say anything. I had no recourse. I said, "Well, I'm going to Adak. Bye." I cried. Boy, I cried. I . . . stayed there [in Adak] until the end of December.

[The boss there in Adak was] kind of a male chauvinist. The only reason they hired me is because he thought I was funny because I offloaded boats. All the processors at that time were tied up on one dock, which is hard to imagine . . . if you know anything about the size of the fleet now. You couldn't tie up all the processors in one line if you had a dock that was a million miles long, there's so many of them. But this was when there were very few of them, and they'd all bring their fishing boats in. They'd come in to deliver, and so after work I'd run and offload each boat. Each night I'd take a different processor and I'd unload. Oh boy. I caught shit from a lot of guys—"What are you doing down there?" "I'm working. What are you doing up there?" On and on and on. See, I never got paid for any of this stuff that I did, because they had their own offloading crews. If you want to help them, that's fine. I could never be part of an unloading crew [then] because I was a woman. It took a long time, but now there are women unloading crews in canneries. Unloading's not that difficult.

I entertained because I offloaded crab boats. He'd bring his gambling buddies out and "Ho! Ho! Ho! A woman unloading crab boats!" I'm not holding it against him; that's [his] problem. I look at it like—I'm here to experience the industry. I'm here to experience what I can do against nature, how long I can stay without sleep. I like to prove to myself my endurance. I want to know how much I can take, how much I can do, how many boxes I can throw. I want to know what my personal limits are, so that someday when I have to tax them to the max, they'll be there. I always go, "Hey, look. I'm not a ladies libber. I just want to do it because I want to do it. It's a personal thing." They always accept that.

Experiencing Fisheries

Christina spent winter quarter, 1973, back at college in Fairbanks, but in the spring she returned to Dutch Harbor. For the next four years she went fishing and worked in fish processing, gradually developing her fisheries skills. For her formal education, she transferred from the University of Alaska to Western Washington University in Bellingham, and during that four-year period she took some more classes. Work in Alaskan fisheries, however, was her priority.

[I stayed in Fairbanks] until spring and went back to Dutch Harbor that spring. I think that was the year I spent eleven months in Dutch Harbor. I hit Alaska Foods. Alaska Foods has always been my basic standpoint in Dutch Harbor, my basic anchor thing. I really should never have gone back [there] after the boss screwed me [out of my money]. I mean any person in her right mind wouldn't have, except that I like the Alaska Foods operation. I mean the physical pleasure of working on a line that was comfortable compared to the smaller processors. A 425-foot boat gives you a lot of room and space to play around with, whereas something like a 225-foot boat is just gross to work on. Each line has its merits, and the Alaska [Foods] was the best operation going. I got to know the people really well. The boss (who screwed me [out of some money] originally) got promoted, so I never had to really associate with him too much.

We were processing shrimp and I processed king crab. For many years all I did was process king crab. I think it wasn't until the winter of '74 or the winter of '75, they started doing tanner crab, which is snow crab, because at that time you could make tons of bucks. King crab was a big thing . . . so I got involved in that and just kept right on going. I met this really neat foreman, and he taught me a lot about king crab quality control, and it really became a fascination. I really like it. I used to think the work was so boring

. . . that if I didn't take some kind of interest in it, I was never going to survive it. The interest became a passion, and now I'm totally hooked. It's so bad that when I go back to the East Coast [to visit my family] . . . I can't wait to get back. There were a lot of good people, a lot of good fishermen who fished for us, and [I] felt really comfortable and knew them really well. But still, I always had this wanderlust thing—like I got to do something different.

During these years Christina often left Dutch Harbor for short periods of time to work in other locales. She worked in a salmon cannery in Alitak for a month, and she spent a few days in a processing plant in Kodiak. Through her connections in the fish processing business, she also had the opportunity to try both crab and salmon fishing.

I worked in a cannery in Alitak in '74 [for a month. It was] a real cannery . . . where they put fish in cans instead of [processing] frozen fish. I only spent a month in that cannery because . . . the owner and I didn't get along too well. It wasn't obvious to the manager. [The owner] just had this attitude about women. They had to be locked up every night at eleven o'clock. He walked up and down the boardwalk to make sure you were in your [bunk]. I went, "Hey, look, I'm not used to this. In Dutch Harbor, it was pretty free." They [had] tried that . . . when I first went there, but it didn't work in Dutch Harbor. But *here* [in Alitak] it's so ingrained . . . in the salmon-cannery philosophy. I couldn't believe it. He [also] felt like, because you were part of his enclave, he could do anything he wanted. "Because you're part of my company, you've got to put out." [When] that happens, I'm going to go. "Hey, that's not where my head is at." I was also pretty homesick. So one day I went, "I'm leaving."

Somewhere in there I met [this] guy. I happened to be in Kodiak on my way to the salmon cannery [in Alitak], and he goes, "You stay with me and fish all summer." I go, "Oh, come on. I got to go to Alitak." [So] when I returned to Kodiak, my friend goes, "Good. Now you stay and cook with me." I went purse seining for the whole summer, for the rest of the summer, and had a tremendous time. I was cook and deckhand. We had two skiff people, one skipper, and me. I ended up running off the deck, which is really great. I started out in the crab business, which is totally male dominated . . . so when I went salmon fishing . . . that was my first experience with a woman doing the part of a man. [But] I was still his girl friend. I wasn't just hired because . . . I knew what I was doing. I learned more and had a better time than I ever had.

[That] winter I went back to Kodiak. I spent a couple of days working for Gulf Processors in Kodiak, which was a really bad trip in itself. That was when I was down to my last forty-seven bucks. [So] I'm trucking around Kodiak in mud up to my knees and feeling really put out. I called my boss, and he sent me the plane fare to get back to Dutch Harbor. Well, he's always been like that. He's always been really, really nice to me . . . maybe because I know about processing and because we've always run a really good operation together. [He's] the boss I [later] worked for at Poseidon Seafoods.

[Sometime after that] one of [the fishermen I had met] goes, "Well, hey. Why don't you go cooking with me for two weeks?" It was in the middle of the springtime, and we're going tanner fishing, and so we went out in this 85-foot boat. The boat had two guys on deck, plus the skipper and me. I had to work on deck. I went out and had a really good time for two weeks. I cooked, did my job. I was seasick for most of the time, because I am not a great sailor. I never have been. But I keep going . . . and they keep saying, "Oh, Tina, you're too Norwegian. You're too stubborn. What is it with you?" [And I was saying], "God, some day I'm going to kick this habit" [of getting seasick]. I've pretty much kicked it now. So I went tanner fishing and learned a lot about tanner and had a good time. Then later on that [following] winter, the same guy came back [and] I went out with him for eight weeks in the Pribilofs, fishing crab, which was really a bummer because we didn't catch much crab. We only got 12,000 crab for six-weeks' work. We spent Christmas out there. We spent New Years out there. Went to Kodiak twice. Just rotten weather. It blew seventy every single day. There were days when I was seasick, and there were days when I wasn't seasick, and we just never figured it out—exactly what it is in my chemistry. But I did my job, and that's . . . all that counts to them. They go, "God, you're some tough lady!" Not tough, just stupid. I'll never forget. The crew is the greatest bunch of guys I've ever been with. They were really supportive.

After four years of processing in Dutch Harbor, Christina went to work for a plant in Akutan. She was soon put in charge of running a crab-processing operation for Poseidon Seafoods.

Poseidon is in Akutan. I had split Alaska Foods and gone to work for Poseidon because my favorite boss was over there. The year [Alaska Foods] sold to National, my direct boss at Alaska had quit and gone to work for Poseidon Seafoods. I really liked him, and I was looking for something different than Dutch Harbor because it was changing so fast, in the sense of becoming very Seattle-oriented. I just didn't want to be in Dutch Harbor anymore. So I split to Akutan, where Poseidon Seafoods has its major processing interest. Plus the fact—Poseidon Seafoods happens to be a pioneer in the fishing-processing end of things. I had to go to work for them because they know all about new technology. I worked for them for two years. Poseidon Seafoods is very fair [with their help]. They always make provisions for plane fares, and if the season ends early, they send you out. As long as you can fulfill your obligations to stay the season. They say "the season, or six months." Usually it's "the season," because they don't do anything but crab. [So it doesn't last six months].

Our boss [at Poseidon] is a real good guy. I worked for him for four years [at Alaska Foods, and he said], "I know what you can do, so come work for me." Poseidon Seafoods . . . have let me run my own processor at one time. I spent last winter . . . hiring my own tanner crew. Going up there and running blue crab, which is a form of king crab, and managing a crew through a two-months' strike for tanner. It was my first real effort. I had been acting . . . [as] . . . purser and assistant foreman . . . always on a line doing things that need to be done—plus doing the bookwork [which] is just a sideline. Processing is something I do all the time, but bookwork was really the administrative end of things my boss let me get into, and I'm very appreciative of [that].

The operation basically entails lifting the crab off the boat in a . . . 1,800-pound brailer. It's brought over to the butcher bin. [The butchers] crack open the crab. They split it in half and they throw it usually, if it's a big plant, into a continuous cooker, which is set for anywhere from twenty-two to twenty-four minutes. [Then] the cooker brings it down to a lower level. It rotates through the cooker shelves in boiling water and then it comes out in a cooling flume that runs around the lower . . . processing area. It comes out on a conveyer belt, or a stainless steel table, or whatever. And there's usually a row of guys standing there. They cut. They separate the legs from the shoulder sections of the crab. The legs are rolled through hydraulic rollers, which squeeze the meat out, into a flume. The flume feeds into the greenbelt. Then the shoulder section is thrown on top of a table, and the guys stand there and blow high-pressured water through each shoulder section, and that [meat] goes out into a flume of rushing water. It rushes down, around, and both . . . small flumes end up on the greenbelt.

Greenbelt is a meat-packing belt. It's a rotating conveyer [belt]. All the crab meat comes down, and women stand on each side and pack it in boxes. To pack a king crab . . . there's a certain design involved, according to certain specifications that the industry has. The person in charge of all this has to be aware and has to instruct everything to make sure you balance the line out, so that you have exactly this . . . much red meat . . . and this much legs.

I *do not* know one guy who could put up with the torture of standing at the greenbelt. Men can't handle the mental strain of tedious work. Women have more endurance. I have run crab lines in excess of twenty-five women. I think it's a credit to women. They have that much control over their minds and their bodies to be able to do it. A lot of plants in this world wouldn't run without women on the line.

I found myself cleaning up after crab season, loading the last freighter out, being in charge of putting pots on the beach—meaning the crab boats come in and take their pots off on our dock and store them on our property—Poseidon Seafood's property. I found myself managing that operation, sending my whole crew out, hiring my whole crew again, going back up there, processing. I thought, "Whoa! This is where I want to be!" But then, they brought in this guy—it's always they bring in the guy—and he's spent twenty days in the crab business, and I'm not kidding you. *Twenty days!* But he had an . . . extremely commanding appearance. He's twenty-eight years old—Annapolis graduate and the whole shot. Our boss . . . when he left the boat, he wouldn't make a decision who was in charge. And it became a big thing between me and [the new guy]. So, I just didn't want to hassle this anymore. I'd proved myself, and I had acted in a very businesslike manner with Poseidon Seafoods, and I felt that I was being screwed. I went, "Well, you guys can just take it and shove it, because I'm going to get out. I'm taking a leave of absence because I can't handle this. You have to make up your mind as to where I stand. I can only do so much." The thing is, I never up to [then] had made a commitment to a company for a long period of time. What I used to do is go out and work really hard and impress a lot of people in a lot of ways. [I decided] to come down to Seattle and spend a few weeks just kicking back.

Christina returned to Seattle on a boat owned by Northern Lights, a freighting and processing company. A couple of weeks after her return to Seattle she went to Northern Lights and asked the owner for a job. She spent the following summer in Alaska processing salmon for Northern Lights.

I came back to Seattle and I went and got a job with Northern Lights. Northern Lights is basically a freighting operation [based in Seattle]. The freighter is called *Neptune,* and they turn it into a salmon processor in the summertime. [The company] is run by fishermen. Because my boss is a fisherman, his attitude is strictly fisherman-oriented—"Sure I'll give you a few cents more a pound, because I know what you go through out there." That can screw you so bad when it comes time to tally up at the end of the season. It's just—it's not a good attitude. [Northern Lights is] a fishing-processing-freighting operation, which is kind of unusual. In general, most processors are processors, and that's it—that's all they do year-round. But my boss, the president of the company, makes the change, and we go to a place called Chignik and process salmon all summer long. Chignik is a lagoon, and there's a very kind of closed fishery. In fact, it has its own permit. It's not just a general area. The people who live there, live there most all of their lives. Very small. We take the *Neptune* to Chignik for boats of small fishermen there. They say, "Oh yeah, we'll fish for you." It's a really small fishery, in

general. We did about a million pounds of salmon this year, which is good
for our size processor. We had twenty-three people on board, of which I was
the only woman . . . who stayed for the whole season.

[This was my] first year for salmon [processing]. We were custom packing.
I packed salmon—making boxes, storing salmon in the freezer—the whole
thing. This [past] summer was a real tough one for me. I physically almost
died this summer . . . because it's been a long time since I did that kind of
agonizing work. I've never worked for a boss that requires you to work thirty
hours [in one go].

[To process salmon] usually there's a tender sitting alongside the boat, which
is a holding boat with chilled, circulating salt water in it. We dump the fish
off the small boats into the holding tank and then, because our particular
capabilities are limited, we just brail the fish off the boat into our line. When
they dump it on the line . . . they usually "princess dress" salmon, which is
[to] take the gills out, and slit the bellies open, and take all the guts out.
Then it goes from the upper deck to the lower deck . . . and it's panned up
in metal pans and frozen. Then it's cracked out of the pans and hand-glazed.
Each fish is individually dipped in a sugar solution and thrown into a sixty-
pound box. Sugar . . . helps keep the glaze intact on the fish. It keeps . . .
the moisture from evaporating when it's frozen. It keeps the glaze from de-
hydrating.

I [also] did all the work on the boat as far as keeping records and stuff.
Northern Lights is so small. Their office staff is one accountant and that's it.
When salmon season [in Chignik] ended, I came down here [with the ac-
countant] and helped clean up for the broker. Custom packing is kind of an
unusual deal, in a way. In Seattle there's a lot of small canneries that do
custom pack. But in general, the industry itself doesn't custom pack if they
can help it, because they're out to make money and custom packing doesn't
always make you money. We're lucky this year. The contracts were made
before the bottom fell out of the salmon market. So I ended up down here
helping [the accountant] out with the broker [confirming the quantity of
packing that was done], and picking up a lot of information on an end I was
not familiar with before. My education in salmon this year [was] really tremendous.

Current Fisheries Interests and Options

*Christina has several options available to her, developed through her past fisheries
contacts. The current man in Christina's life, to whom she refers as Number One,
owns a 155-foot fishing processor, the* Alaska Dawn. *It is a dragger, suitable
primarily for catching bottomfish. Christina would like to run the processing operation on the* Alaska Dawn. *She has been helping Number One set up the processing line on the boat, and she has also succeeded in selling the* Alaska Dawn's

salmon, scallop, and crab inventory. This was her first venture into yet another
aspect of fisheries: marketing.

I am probably the sole representative of the sales department of *Alaska Dawn*
fisheries. I'm not paid because I'm a really good friend of the boss of the company. His whole salmon inventory [was] sitting up in Bellingham. The salmon
[was] killing them in the marketplace because they didn't princess-dress their
salmon. [But] I finally . . . sold the salmon!

Bottomfish is something I know very little, if anything, about. I went to
Fish Expo, because I wanted to check out quality control. I wanted to see
what was happening. *Alaska Dawn* is strictly a dragger. She is really geared
for dragging, and he's really going to hit it hard on bottomfish. The bottom
will fall out [of the crab business]. The crab business has always peaked in
about two years and fallen flat out, and then peaked again. It's fallen down
[now]; it's really fallen. Sooner or later the canneries are going to have to
admit defeat, since none of them are rigged for bottomfish. Not one processor in Alaska is rigged for bottomfish exactly. That's what he wants to do
[and], like I say, *Alaska Dawn* is my passion. I want to see it work more than
any boat in the industry. I want to see him be able [to] stand up to all these
processors and say, "I did it. I did it." I really feel that with his fishing expertise and my processing expertise, we could do it together. [Number One]
let me build two greenbelts. I really loved that—designed them . . . at Sound
Equipment in Bellingham. If I go back to *Alaska Dawn*, if we go bottomfishing and do herring—he keeps saying, "You're going to work for me? You're
going to work for me?" And [I] go, "Ah, you're such an asshole. I don't know."
But, if I do, I'll be in charge of all the processing for *Alaska Dawn*.

Although she would prefer to work with Number One, Christina is also con-
sidering other fisheries options.

Right now . . . I'm really seriously considering going to Seward. [My boss
at Northern Lights is] going to sail with the *Neptune,* and he wants me to go
up with him and cook on the *Ocean Traveler,* a fishing boat [of which he
owns part] that he's going to run. He's trying to put together a raw tanner
pack, because raw crab is an interesting thing on the Japanese market. [If]
you get it, it's pretty lucrative. Right now it's more per pound than cooked
crab because it's a specialty item. So what he wants to do is take the *Neptune*
up, lay it in Seward, and then fish [for crab] with his fishing boat. He wants
[me] to cook with them, but it all depends on whether or not . . . he gets
the tanner contract. He doesn't know right now if he's going to get it or not,
or if it's going to be worth his while. [That would last from now, January,]
probably till herring season, which is around April 15, April 19. Then he
would take the *Poseidon* to Togiak to freeze herring, and *Ocean Traveler* would

go to Togiak and tender herring back to Seward. After herring season, it'd
be salmon in Chignik again, and it would be [the] same old cycle.

*Poseidon Seafoods, from which Christina took a leave of absence, has asked her
to continue working for them.*

I was offered a really good job the other day to go back and run a herring
line on the *Aeolian*, for Poseidon Seafoods. I got a really good shot at running this herring thing, which is interesting, because I don't know a thing
about herring. But I can run people, and that's probably what you're paid for
these days more than anything. Because it's very difficult to get people to do
their jobs. It just is. Especially because the processing business . . . exists on
its [transient] population. Plus the age strata is eighteen to twenty-two. And
they're, "Oh, I'm out to see Alaska. Wow!" It's not necessarily, "I'm out to
work." And it's very difficult, when you're caught in the middle of a big run,
if you've got a crew that goes, "I don't want to do this. I don't want to do
this." [I go], "God! Get it together, C'mon, we've got to do it!"

Plus, wages are—as far as I'm concerned, personally, it's high-paid slave
labor and that's it. Because, four bucks an hour, even though with room and
board—it's bad. It's . . . not good. Not for what you're required to do. Then
again, *Alaska Dawn* pays six and nine [six for the first forty hours, nine for
overtime], and they couldn't get any more out of their people than people at
four and six. The crab processors . . . are going to have to offer their crews
guarantees and a whole different scheme of payment and stuff. They're usually under a six-month's contract. People get screwed because there's not
enough fish and they're stuck there. Right now the quota for tanner crab is
27-million pounds, and if you have any idea, 27-million is about the capacity
for 50 boats—about one trip for 50 boats. There are 250 boats in this thing.
If I decide to go with Poseidon Seafoods, I will be senior foreman with a shot
at being a superintendent. I will be senior foreman in charge of the *Aeolian*,
which is a 365-foot boat.

[Those] are good options. They could pay off very well. But see, I never
looked at the industry to pay off, because I always figured, people on the
line—you think of yourself as a lineworker making four bucks an hour. Big
deal. Never going to pay off. The people who get paid off are the fishermen
on deck. I've just never felt like it was a big money thing. It's really interesting. But now, I'm kind of looking at it in terms of money . . . because
. . . I've definitely committed myself to the industry, and [now] I want to
make as good a living as possible. I'd like to see them pay me for what I
know. But the thing is . . . if you pick [small] operations like Northern Lights
or Alaska Dawn, you never know where your next buck is coming from.
Technically, it's the way the whole company's run. You don't know if this

minute, if the bottom falls out of the market, you're going to be left with no money, or you're going to have all kinds of money. I like it.

As far as what I would like to do *careerwise*, I would like, yes, to be super-intendent of my own processor for, say a high-pressure king crab season and tanner crab season, because those are the things I feel most familiar with.

The fish business . . . is the greatest industry in the world for people who want to work and people who want to go places. Go *high* and make lots of money, because there *is* no talent. It's all very tentative. If someone really wants to apply themselves, they can become superintendent. There's just *not* a lot of qualified people. It doesn't really require any specific education. A few guys I know have gone places, and they've got degrees in fisheries biology or were biologists to begin with. A few are business administration people. It's a great opportunity for somebody who wants to go out and grab it, be it man or woman.

Although Christina sees many opportunities in fisheries for those who wish to pursue them, there are certain aspects of the business that continue to be less accessible to women. Crab fishing is probably the most difficult activity for women to get into.

If you want to talk of the crab business and women . . . they're pretty nonexistent, unless they're . . . a skipper's wife or girl friend. Women in the crab [fishing] business . . . are really few and far between. It seems like the crab business has the corner on male ego, *extreme* male ego. I think [it's] be-cause of lack of education. They're workingmen. They have workingmen's attitudes; typical blue-collar-worker attitude. These guys are—I don't know if it's . . . insanity, or out to prove what, but some of them are exceedingly tough. They have no soft streak out there. But then, to be a successful high-liner, you have to have that. I'm waiting for the day there's a lady crab skip-per. She'll have to be *as* tough, if not five times tougher. And is it worth it? It's not ever going to be prestige in the sense of a man. It's going to be more like notoriety.

Women are usually limited by their lack of physical strength . . . on the big crab boats . . . in excess of 120 feet. A woman really . . . cannot do the job on deck, because physically handling a six- to eight-hundred-pound pot is just not the easiest thing in the world, even with . . . sophisticated equipment. Women can only lift so much; my max is about 114 pounds, deadweight . . . and that's really pressing it. The guys . . . can't stand it, either, but they're of course physically stronger . . . [although] women have more endurance.

But . . . in a small boat fishery . . . like Kodiak or like Southeastern, you find a lot more participation by women, because they're usually two-man operations [or] a man-woman operation—husband-wife, girl friend–boy-friend, father-daughter . . . that sort of thing. But in small boat operations, you also have a better chance as just a woman. [So if you] go check with this or that skipper, you can get a job.

What I know of the American fishing-processing effort, usually both boats are too small to have the legal facilities required for women. They say, if you're going to have women aboard, you have to have separate facilities. And usually the boats are just too small. God! The Japanese . . . and Russians have 500-foot mother ships. [The United States] biggest fishing-processor is the *American Number One*, 165 feet.

A woman can say "I want a job" [on a fishing-processor] and go sue someone because they couldn't give her a job. But what women don't realize in the fish business is that they're stepping on other women. Wives play such a tremendous role in fishermen's attitudes towards women. I know a lot of skippers who are good guys . . . who would be willing to give a woman a shot [except for the] hassle involved in explaining that to your wife! "No, dear, I'm *not* sleeping with her." You have to put yourself in their place. It is a mental hassle. A husband . . . stays away for six, eight months out of the year. Here she is sitting down here. I've never figured out a decent way to overcome this yet, except when I go aboard a boat, I will wear my baggiest dungarees, my baggiest shirt, and I will . . . adopt a manner that is not exactly feminine. [Then] if his old lady walked aboard the boat . . . she'd look at me, and she'd go, "God, I don't have anything to worry about!"

Personally Speaking

[Life in the fish business], it's like a circle. It's cyclical. [A few months] is the longest period of time I've ever spent in Seattle. But up until this point I'm . . . flitting back and forth [to Alaska]. These things are momentary. I could get a phone call tonight [from my boss] that says, "Well, I'm leaving tomorrow, Tina. Do you want to come?" "Well, God . . . I don't know. Let me see. How long does it take me to pack my stuff?" One duffle bag [is] on the *Ocean Traveler*, and one duffle bag is in the back of my car, packed and ready to go. You never really establish any real roots.

Most of my friends are involved in the fish business in many ways. Who do I know outside the fish business? Right now in Seattle? Everyone [I know] is involved in the fish business in some way.

I'm really of an ambiguous nature. But I really want to see who I consider my man succeed. Yes, and the *Alaska Dawn*. But yet, to see him succeed, I have gone out and been my own person in the industry. I have learned what I've learned and . . . I'm still learning. Every day I learn something new about either the marketing end or something. I am my own person, definitely. But yet, in a sense, it's all been geared towards knowing everything so I can work towards something for him. I'm not exactly involved in the sense of being a wife and wanting to see my husband succeed. It's not just self-interest [either].

I've been going to church regularly since I've been down here. I really find a lot of comfort in Mass right now. I was never brought up strict anything [but] we have two of these swinging priests over here. I think they're pretty liberal, actually. Anyway, I enjoy going and listening to what they have to say. I don't know why [but] I'm feeling quite guilty about my past personal life. Really, I shouldn't. [My going to church is] kind of weird to a lot of friends of mine who are straight out of Alaska. They just go, "But that's not you!" It's a part of me.

Basically, my father's whole problem was—and I saw it after I went out and met the world on its own terms—is that nothing in this world does come free. Idealism is beautiful and it's great, but it's a luxury that few can really afford. To stay idealistic forever, you do have to temper your idealism with realism and . . . you have to open your eyes. [I] have learned that there is a happy medium. [I've learned also] that there are people who are out to screw you. And, yes, work is not fun sometimes and, yes, you do have to work hard to get to where you want to go. I do understand him more now, and I always want to come home and say, "Hey Dad, I really understand where you were coming from now. I really understand why you were so upset . . . because you thought I was going to waste my life studying sociology and anthro."

So, that's [my] life history. I was born and fell in love about a zillion times, and I haven't died yet but I'm planning on it, definitely. I think everyone has to plan on it, at least physically. Probably get reborn again. Only next time I'll probably be into some Norwegian fisherman's body.

Conclusion

When we began our study of women's roles in Pacific Northwest fishing, we had little to guide us in our inquiries. No documentation of women in Northwest fishing existed except for a few popular accounts and newspaper articles. The present volume, containing stories of ten women's lives told in context of the socioeconomic and political history of the Pacific Northwest fishing industry, is the first attempt by anthropologists to document the range of choices available to women who wish to participate in this industry.

In telling the story of women in this occupational community in the life-history format, we have drawn not only upon women's tape-recorded life histories but also on topical interviews with members of the larger fishing community, observations at formal and informal meetings of commercial fishing organizations, results of the Fish Expo questionnaire, and miscellaneous sources available regarding the industry. The limitations of the life-history methodology are numerous, but most obvious is its reliance on human memory. As we have indicated in our introduction, people selectively tell the stories of their lives to emphasize political and personal perspectives, or to provide good images of the lives they have lived. Thus, researchers may question the reliability of information obtained from oral histories when they are used as the only source of information. Contextual information from a variety of sources may offset doubts. By employing a range of research techniques, we have been able to bring together sufficient information to make several generalizations about these ten women and their work in the Pacific Northwest fishing industry.

We cannot generalize about all women and their statuses or about all women in the Pacific Northwest fishing industry on the basis of what we have presented here, but our analysis may suggest some ideas useful in future research. For example, we offer some generalizations about the patterns of women's work and other involvements in the Pacific Northwest fishing industry, recognizing at the outset that the ten women whose lives are presented here represent diversity in terms of: ethnicity; age; marital status; their own upbringing and the child rearing they provide; other personal and familial responsibilities; length of time of involvement in the industry; and occupational choices or roles within the industry.

The women whose life histories are included in this book are involved in various sectors of the fishing industry. We can summarize their work activities and discuss their participation in the industry in terms of several broad categories: fishermen's wives; women in small family businesses and independent women; fish processing; fishing industry management; and political activism. These categories should be viewed as situational and nonexclusive: women may participate in the industry in several capacities sequentially or during any given time period.

In much of the social science literature, women's work activities are often discussed in terms of whether they are *traditional* or *nontraditional*. We have chosen to use the words *customary* and *noncustomary*.[1] Defining what constitutes customary work for women is a complex task, as what is considered customary or noncustomary work for women rests not only on the nature of the tasks performed but also on the social and physical contexts in which those tasks are embedded. The lives of ten women in this book illustrate this complexity. Their involvement in the fishing industry ranges along a continuum from what is viewed as expected work in women's domains to what is seen as work in men's domains, both in terms of the types of tasks and the physical settings in which they are performed. In addition, whether the work performed by the women is viewed as customary or noncustomary appears to depend to some extent on the social structural relationships within which the work is performed. We have summarized the ten women's work activities under the following nonexclusive categories: fishermen's wives; women in small family businesses and independent women; fish processing; women in fishing industry management and political activism.

Fishermen's Wives

In the stereotypic view of the fishing family, men fish and women stay at home to take care of children and home work. The domains of wife and husband are viewed as entirely separate. Only recently have researchers begun to note that fishermen's wives contribute to the industry even though they may not participate in the harvesting of fish (see Danowski 1980, Ellis 1984, Thompson 1985). For example, because fishermen must travel to the locale where their targeted species is available, a fisherman's wife who does not participate in the harvest assumes the primary responsibility for maintaining the family home (including finances) and for child bearing. While her husband is away, the fisherman's wife may be able to call on others in similar situa-

---

1. The set of terms *traditional/nontraditional* implies greater antiquity in transmission of cultural practices. Whereas the terms *customary/noncustomary* represent "a group pattern of habitual activity, usually transmitted from one generation to another," *tradition* is "the handing down of statements, beliefs, legends, customs, etc., from generation to generation, especially by word of mouth or by practice" (*Random House Dictionary*, 1967). The main difference between these two sets of concepts is that *customary* implies a shorter length of time than does *traditional*.

tions for help, but as Tink Mosness has pointed out, in emergencies she relies on herself.

We interviewed several women who participate or have participated in the industry as fishermen's wives. Gladys Olsen and her husband maintained very separate domains of action, and the circumstances of her life give added weight to the words "primary responsibility." Gladys lived alone with her children on an island during the early years of her marriage. Her husband, who was employed by a cannery and worked on a tender in the summer and with the watchman during the winter, came home when he could. She says, "He was home more often during the winter, at least once a week." Gladys grew her own vegetables and "had food from the beach" in addition to the provisions her husband brought home from the cannery stores. When her husband became a salmon and crab fisherman, she saw even less of him. Gladys was responsible for running her household unaided. She says, "There was a time, after I had moved [back] to the village [of Afognak], where [Hans] felt that I should learn to take care of everything myself. Just made no difference if it was a man's job. . . . And I did. I did." Gladys and her husband had seven sons and one daughter; five of the sons took up fishing either full or part time. Her husband did not believe in taking their daughter fishing, although in later years Gladys and her daughter went fishing with her husband in the spring to obtain fish for "home use."

While Evie Hansen was growing up, it appears that her mother's living situation was similar to that of Gladys Olsen. Her father was gone fishing for most of the year, and her mother took care of the family and also ran their chicken farm. Evie says of her father, "He was from another type of life-style where the men didn't do anything in the house. They brought home the bacon and from there it was for the women to do everything about the bacon." Evie's family grew and preserved their own vegetables and raised animals for food. Evie's father took his two sons fishing, but not his daughter. One of the sons became a fisherman, while Evie entered fishing through marriage to a fisherman.

The lives of two other fishermen's wives in this volume show variations on the theme of separate domains for wife and husband, and the degree to which wives or husbands participate in what they perceive to be one another's domains. Tink Mosness's husband fishes in Alaska each year starting around April 1 and then returns to fish in the San Juan Islands during the summer. During her husband's early years in fishing, the family lived close to his summer fishing grounds; later the family moved to the outskirts of Seattle. Every August the family rented a cabin directly on her husband's fishing grounds. She says, "We rented it for eighteen summers. . . . We had it for the month of August. Pete would come in in the morning and anchor his boat right there." In later years, she would go fishing on Puget Sound with her husband—as a companion, not as a worker. She says, "Sometimes [my presence] is accepted with gratitude and sometimes it does me more good than it does the fishing part of the family." Tink Mosness and her husband have three daughters, one of whom has become a fisherman.

Thus, while it appears that Gladys Olsen occasionally fished with her husband, either for recreation or to obtain fish for the family, she never went fishing with her husband during the main fishing seasons. Tink Mosness, however, went fishing during the main season as a companion for her husband, and Evie Hansen went fishing as both a companion and a worker.

Small Family Businesses and Independent Women

Some fishermen's wives are primary participants in a small family business. It is customary for a wife and husband to work together to ensure the success of such a business, especially when the living and work spaces coincide. Where the two spaces do not coincide, the wife's contribution may go unnoticed. Many fishermen's wives participate in the family business by doing bookkeeping, helping with gear and boat maintenance (such as net mending or boat painting), and running errands (such as obtaining replacement parts). That this is a well-established pattern is confirmed by Tink Mosness's account of how boat parts were used as a code to inform the Coast Guard of foreign fishing activity. Fishing husbands passed coded information on the size of a foreign vessel, its country of origin, and its location to their land-based wives, who then called the Coast Guard—all without arousing the suspicion of the target vessels. Wives who are primarily land-based may also sporadically move into their husbands' work space. The wives may serve as companions or occasional emergency replacement-crew members, or they may work as crew members more or less regularly. For example, as noted above, Evie Hansen gillnetted with her husband, except when the water was very rough.

Sometimes the families decide to combine their living and work spaces and move their families onto their boats. Helen Gau and her husband chose this option beginning with their second fishing season. They and their three children all participated in harvest tasks. Her husband was captain, did mechanical work, ran the fishing gear, and iced down the fish. She provided back-up for all these tasks except the mechanical work. She says, "As far as running the boat, of course, I can handle anything that comes up." She cleaned fish, cooked, and performed other domestic tasks. Their two eldest daughters helped by watching the baby and doing some of the cooking. Later, by their own choice, the children participated in such fishing tasks as baiting hooks and cleaning fish. When the children wanted actually to catch the fish, Helen ran the gear and helped the children land them. She says of her children, "[They] were certainly good crew." Although Helen Gau was living under conditions and performing tasks usually not considered customary for women, she was still involved in a customary family relationship. In her role as crew member, she participated directly in the family business while her husband remained head of the family, the business, and the boat.

Mars Jones also began her fishing career by assisting in a family business. At first, she and her husband had a small boat with which they fished along the beaches. Later, they operated a larger boat and began chartering. She assisted her husband for three years. Then, a man asked her to run his boat, but she needed a license. When she applied for her license and was asked why she wanted it, she explained that not only did her husband sometimes get migraines but that she also didn't like to turn people down. "I wanted to be sure that I was doing the right thing by running the trips, and I wanted my license." She ran three or four different vessels for about five years and then obtained her own boat. She runs her own charters both summer and winter, and her husband trolls during the summer and charters during the winter. She and her husband are both involved in the same business yet run separate operations. Both are captains of their own boat. Mars Jones does her own boat maintenance, with the exception of engine work. Each lives on her or his own boat during the summer. When the season is over, they return to a more customary domestic arrangement, living in Port Angeles and running daily charter trips.

By running separate operations, Mars Jones and her husband avoid one of the pitfalls sometimes encountered by married yet independent women fishermen. The usual line of authority on a fishing vessel is vertical, from the captain to each of the crew. In a customary view of the American family, the husband is the head of the household and has authority over the wife. If the wife is also a crew member, these two vertical lines of assumed authority coincide. If she is the captain and the husband is the crew member, the two lines of customary authority conflict.

In her fishing days, Lois Engelson was an independent worker who always ran her own operation. Multiple attempts to combine wife-husband and captain-crew relationships failed. Her second husband could not deal with her independence and wanted her to give up fishing. She fished a few seasons with her third husband, and during this marriage, for a few seasons she fished alone on a leased boat. She says of her third husband, "He went fishing with me, but he couldn't handle how involved I was in fishing." Toward the end of her fishing career, she married her fourth husband. Although he went fishing with her, her attempts to interest him in becoming a full crew member failed. She says, "[My husband] did real fine on the boat for handling gear—he did excellent. And he does fish—I showed him how. But he would never run the boat."

Marya Moses, another independent fisherman, made no attempt to combine marriage and fishing. Her husband had died some years before her involvement in fishing. When she started fishing, she was captain and her daughters were her crew. Another relative worked on shore as bull wincher. Here the captain-crew and mother-daughter lines of authority coincided. In later years, her grandchildren worked as her crew. She says, "The crew that I pick, they're handpicked. They have to know that they're doing. They've got to take orders." As grandmother, Marya Moses perhaps has greater authority than would a younger woman. Linda Jones points out that "Indian society is a little different. My grandmother is the head of our family. . . . She gives the orders and the people obey out of respect." Nevertheless, from other of Linda Jones's comments, it seems likely that if a Tulalip wife and husband began fishing, the husband would expect to be captain.

Thus, Mars Jones and her husband avoid conflicting lines of authority by running separate operations. Marya Moses had no husband in fishing. Her choice of crew coincided with lines of authority within the family. In Lois Engelson's case, conflicting lines of customary authority probably added stress to the husband-wife relationship.

The problem of coinciding or conflicting lines of customary authority also influences the position of independent women who work for crew shares or wages. During our research, project personnel talked with male fishermen who stated that, in their view, having one woman on board among several men causes disruption. One circumstance in which this might happen would be if a female crew member became sexually involved with a male crew member. They would become strong allies whose primary allegiance would most likely be to each other. This would introduce a strong horizontal bond possibly conflicting with the vertical line of authority from captain to each individual crew member. Project members heard of instances where a lone woman among a crew of three or four was the captain's "significant other." Christina Jefferson, for example, worked one summer as a cook and deckhand on a boyfriend's seiner. She also worked as a cook and deckhand on a Bering Sea crab boat; but that, she says, was "just a job." We have not heard of instances where a female and male crew member became romantically involved and remained fishing as part of the same crew.

Of all the women's roles and tasks discussed so far, those of land-based fisherman's wife, whether involved in her husband's business or not, are usually considered customary activities. Wives who work as crew members on their husband's vessels are fulfilling the customary roles of helpers (and sometimes partners) in the family business; but they are also doing noncustomary tasks in noncustomary settings. In family businesses, the work place and the home place often coincide, but are not usually located at sea. Mars Jones's situation presents a variation on these themes: she is captain of her own vessel but is, in effect, the manager of a branch of the family business.

Independent women who fish also perform noncustomary tasks in noncustomary settings, but they deviate from the customary in yet another way.

They are neither helpers nor back-up personnel; instead, they are independent, self-reliant, and "in charge" in a rigorous environment long considered to be the exclusive domain of men. Women who are independent wage workers or who work for a share of the harvest may be doing the same work on board that a skipper's wife or girl friend would do, but their independence makes their social position noncustomary in every respect.

Other women workers, although independent, may be performing tasks which are extensions of home work, and therefore their work may be viewed as customary. Such is the case with fish processing.

Women in Fish Processing

Women have been customarily responsible for preparing, preserving, and cooking food. They have been involved in fish processing for wages in the Pacific Northwest (and elsewhere) since at least the late 1800s. Their job is to turn the harvested fish into a marketable (or storable) product. Although the tasks involved in cleaning fish can be viewed as an extension of home work, the physical and social settings in which such work is done may vary considerably.

Many processing operations in Alaska are in isolated areas or are on ships rather than on land. Christina Jefferson's first fish-processing job was on a Dutch Harbor Liberty ship. A friend's father did not consider Dutch Harbor an appropriate place for his daughter to work. Christina Jefferson says, "We had gone down and applied [for Seldovia] and we got sent applications for Dutch Harbor. Well, her father didn't want her to go [to Dutch Harbor] . . . because it was a whole different atmosphere than Seldovia, which is a small boat fishery." In addition, many of the plants in these settings may have much larger numbers of male than of female workers, and the lives of women workers on and off the job may be restricted. Christina Jefferson mentions the "salmon-cannery philosophy" of an employer who imposed a nighttime curfew for women workers. Such curfews represent an attempt to control the movement of women in settings considered inappropriate for them and constitute a holdover from a previous era.

Although Christina Jefferson was performing tasks usually thought of as women's customary work (i.e., working on a processing line with other women), some of the social and physical settings in which she worked were not considered entirely appropriate for women. She was, for example, cleaning fish in a cold, wet, and sometimes dangerous environment. In addition, she developed a very migratory life-style, moving from place to place with the fishing seasons for the various species. Later Christina Jefferson became a line supervisor, but she had trouble moving into management, a male-dominated area in fish processing.

Women in Fishing Industry Management

Marti Castle, a processing plant manager, sums up Christina Jefferson's difficulties in moving into management when she says, "Women . . . in the fish business, they're not that common, except as traditional secretaries or bookkeepers. In Alaska [even] most of the bookkeepers in canneries are men. They're almost always men. A lot of your office staff are men. Women will be floor ladies or kind of a floor foreman. That's a pretty common spot for a woman."

A woman in the position of "floor lady" uses management skills to accomplish a set of tasks. Many writers have pointed out that, since housewives manage their own time and tasks as well as children and the tasks of others within their households, management may be thought of as an extension of home work (see Kanter 1977). It is not customary, however, for women who manage their households to direct the activities (at least overtly) of adult men. Moreover, in many communities in the United States, it is assumed (correctly or incorrectly) that a wife is an agent carrying out a husband's overall directives. As floor lady, Christina Jefferson directs the activities of other women and any men who work on the lines. But movement into the middle and upper levels of management is difficult. At those levels, managers have both the openly acknowledged authority and the structured responsibility to direct the activities of others. In most cases, these "others" include adult men. In the case of women who wish to move into management, the issue of who is competent enough to exercise authority over whom often becomes obscured by other individual beliefs and sociocultural conventions.

Both Marti Castle in fish processing and Linda Jones in tribal fisheries management ran into the assumption that a woman cannot know as much as a man. Linda Jones says, "They don't think of you as being a competent individual first. They think of you as being a woman and therefore a little weaker, a little more emotional, and probably not as capable as they are." Marti Castle has also encountered resentment from male workers. She says, "There is a lot of resentment from men working under a woman, often. I've seen that a lot. . . . Sometimes they don't want to be told anything by a woman. And other people, they could care less."

Linda Jones, living and working in a small Indian society, had three other factors with which to contend: age, marital status, and the politics involved when one works in a multitribal context. Regarding her relative youth, she says "There's always been traditionally a great deal of respect for the tribal elders, for the older people. Consequently, having a young person come in and give orders and . . . decide when and where you'll fish and when you won't is real hard for some of the older guys to swallow." She also believes that her single status has been a disadvantage in her role as a tribal authority. She says that another Indian woman with a position similar to hers "is accorded a certain amount of respect because she's married and she has a family."

Marti Castle did not seem to encounter problems because of her age, and problems deriving from her marital status were somewhat different from Linda Jones's. Like many other women in fisheries, in relationships within the workplace, she never lets "that sexual edge sort of slide in." As one woman fisherman stated, "You put your sexuality in your back pocket." By doing so, expectations based on knowledge and behavior appropriate to an occupational position in a work structure take precedence over expectations based on assumptions about female-male relationships and the "nature of women." This kind of compartmentalization of expectations, however, may be more easily accomplished in the wider society than in a small, encapsulated, traditional community, especially when the managerial position involves negotiating conflicting political expectations.

Linda Jones had to balance the desires of members of her own community with those of other tribal groups in the context of legal parameters set down by the Boldt Decision. As Director of Fisheries for the Tulalip Tribes she not only managed her own tribes' fishing effort but also represented eight additional tribes in her roles as chair of the Point Elliott Treaty Council and alternate commissioner to the Northwest Indian Fisheries Commission.

Women as Political Activists and in Fishing-Industry Politics

Politics is an integral aspect of commercial fishing in the Pacific Northwest. A number of different user groups on both the international and regional levels often compete with one another in their efforts to gain some measure of control over what is, at times, a scarce resource. Significant regional user groups include recreational, nontreaty, commercial, and treaty fishermen. Women and men representing various interests in species available for commercial and recreational harvest have worked alone or in coalitions in attempts to win a greater allocation of fish for their respective interest groups.

Tink Mosness and Evie Hansen have actively participated in the politics of the commercial fishing industry. Tink Mosness has been a motivating force in founding several voluntary political action organizations devoted to furthering the interests of commercial fishermen. The first organization to which she belonged was the Puget Sound Gillnetters Association (PSGA) auxiliary. In 1954 the members of the PSGA asked their wives' help in distributing informational literature aimed at defeating a state initiative that would have closed Puget Sound to all but recreational fishermen. Tink Mosness says, "After the election the fellows asked us to come back to the meeting. And we expected them to thank us for our support and kind of give us a rah-rah-rah for what we had done. But instead, they said that from that time on . . . they had voted that we were a part of their organization. So they sent us in the other room to form an auxiliary."

Although the formation of this first women's group came about through the prompting of their husbands, PSGA auxiliary members quickly became autonomous and influential actors in the political arenas concerned with fishing. Evie Hansen, who joined the auxiliary in 1974 in response to the Boldt Decision says, "We've had some run-ins where they [the men's group] haven't liked what we've done, and we've come on without them. That's their problem, not ours. We haven't liked what they've done, either. Yet our goals are the same; there's really no conflict as far as our goals are concerned."

At first, the PSGA auxiliary concentrated on increasing public awareness of the fishing industry and on increasing consumer demand for the fishermen's products. They sponsored fishermen's floats in local parades and gave out free recipes in stores. This emphasis on public relations has continued, even as the group has become involved in other activities. For example, in 1978, they published a cookbook containing recipes for many species of fish. PSGA auxiliary members have produced their own newsletter and have edited and distributed the men's newsletter. Both newsletters are sent out on a monthly basis in order to maintain a flow of information among members. In addition they have helped found several other women's organizations involving women who identified themselves as halibut wives, troller wives, and seine wives. They have also been influential in the formation of a series of larger organizations with membership open to both women and men.

In 1963 Tink Mosness helped found National Fishermen and Wives (NFW). Membership in this organization could be obtained through membership in a participating organization or on an individual basis. In 1969 the NFW merged with another fishermen's organization to form the National Federation of Fishermen (NFF). The NFF became an umbrella organization for regional fishermen's groups. As the NFF grew, it added East Coast fishermen's organizations to its membership and established a national office in Washington, D.C. With the formation of the NFF, the members of the PSGA auxiliary and other women's groups entered the national political arena. For example, NFF women became involved in lobbying the federal government to enact a 200-mile limit to curb harassment of U.S. fishermen by foreign fishing fleets. When the NFF did not support Puget Sound area fishermen in the controversy surrounding the Boldt Decision, many Puget Sound organizations withdrew from the organization. The Washington Association for Fisheries, a statewide organization open to individual membership, was subsequently founded and remains active today in both state and federal issues.

The women's groups have continued to exist along with the larger mixed-membership organizations, and new women's organizations have formed. The Pacific Coast Coalition of Fishermen's Wives, for example, was founded in

1976 and consisted (in 1979) of at least fifteen women's groups from the states of California, Oregon, and Washington. The level of activity of these women's groups appears to depend on whether there are events requiring their attention. In 1979 Evie Hansen said, "Looking at the history of the auxiliary, when things have needed to be taken care of politically, the group has been active. We've had four or five years of intensity, and now the auxiliary is kind of backing off."

The groups discussed above (and more deeply in the life histories of Tink Mosness and Evie Hansen) were formed, and women became active in them to protect their families' economic interests and way of life. Greater autonomy and involvement in ever larger arenas of social and political action were and are by-products of this participation. In fact, the history of their involvement parallels that described by Karen Blair for women in late nineteenth and early twentieth century women's clubs (Blair 1980). The fisheries organizations, like the early women's clubs, "served as a vehicle of entry into the main stream of public affairs" (Baxter, in Blair 1980:xi). The work involved in defining, meeting, and carrying forward business objectives is part of the political process used by individuals and collectivities in urban industrial societies based on the international market economy. Such public and political activity is usually associated with men in United States society. The activity of women in commercial fishing organizations illustrates once again that such activities are not restricted to men, and that women are critical actors in this country's sociopolitical organization as well as in the Pacific Northwest economy.

Concluding Remarks

Women's contributions to United States society in general and to the fishing industry in particular are only beginning to be recognized and documented. Commercial fishing has long been an important component of the Pacific Northwest economy, but women's participation in the industry has largely gone unnoticed. Most Northwesterners have been aware primarily of the role of fisherman's wife, but few outside of the producers' sector of the industry have realized just what this entails. Through the life histories in this volume, members of the wider community can become acquainted with the activities of women contributing to the industry in a variety of capacities, and in a broad sense, they can experience the remarkable quality of these women's lives.

The women's stories document activities and problems common to daily life in the fishing industry as well as some of the adventures of its participants. The women have told their stories in the context of the multiplicity of issues involved in managing fisheries resources, and their lives illustrate the ways in which women and men in the fishing industry attempt to cope with and influence continuing conflict and changing conditions.

Postscript
Katherine (Tink) Mosness

The fishing industry has always played a vital part in the history of the Northwest, and in one way or another, women have always been a part of it, but not always recognized. Time changes many things, and some things change for the better. Women are now in all phases, from skippering to owning their own seafood businesses, and the most important part is that they have been accepted in all of these aspects. The first time I went back to New Bedford, Massachusetts, to speak at a newly organized group of wives of trawlers, I was greatly interested in the difference in their vessels compared with the vessels on the West Coast. It was spelled out to me very firmly that I could talk to the fishermen on the dock, I could meet with them, which I did when I spoke to their group also, but I could NOT go aboard their vessels as it could bring them bad luck and bad luck to the vessel. Bad luck!

The salmon fisheries of the state of Washington are still in transition. The Boldt Decision has changed the management policies of our natural resources of fish and water. We still have an influx of foreign products, greatly decreased fishing time, and increased costs of licenses for boats and gear. But a number of good changes have taken place also. Seafood is now a household word. The recognition of the importance of fish and seafood products in our daily diet has increased the demand for seafood products beyond our widest expectations of even a few years ago. New jobs have also been opened up in seafood promotion—care of fish and its ease of preparation. And the American public has caught onto the versatility of the many species. The people who have been doing taste-testing in supermarkets, giving out free samples, recipes, being on TV programs, and promoting recipes have also contributed to this new phenomenon.

The Washington Association for Fisheries (WAF) remains a viable voice for Washington State fishermen and continues to attack problems that regularly surface on both the state and federal levels. Just to maintain the fisheries we now have, we must be on constant guard. We are serving on the newly organized Seafood Advisory Committee, which was formed as an implementation of Washington State House Bill 1362 for a combined industry and government endeavor to promote Washington State fishery products. This has always been one of our more positive goals.

Over the years, fishing has been a good way of life. Many of our friends are gone now; many are retired; and some, like my husband, have just finished another new boat. No, the fishing in Puget Sound does not warrant the expenditure of the money it costs to build a new fiberglass gillnetter. But such an investment is a salute to the future. The fleet is upgrading: new gear, new equipment, and new electronics make it easier and safer to earn a living in the fishing industry. Not all fisheries are safe. We still lose too many fishermen each year. Safety courses, mandatory regulations for the larger vessels, and more safety equipment on board are helping, but not fast enough.

In 1985 twelve of us formed The Seattle Fishermen's Memorial Committee. We have raised over $80,000 towards our goal of $100,000. When Fishermen's Terminal is reborn after renovation, its centerpiece will be a twenty-eight-foot bronze and stone memorial erected as a tribute to the men and women in the Seattle fleet who have lost their lives in the pursuit of their fishing livelihood from 1900 to the present. The memorial is being created by Seattle sculptor Ron Petty, whose design won out over seventy others in a competition held by the committee. The finished base section of the statue [was] on display at Fish Expo October 28–31 [1987]. Seattle is a large fishing port and has never had a memorial inscribed with all of the names of those who have lost their lives. We are going back over the records since 1900 to find all of these names. The immense job of compiling names to be memorialized has given us a deep appreciation of the immense size of this industry, and the interlocking nature of families, and the feeling of community among members of this industry.

As I reread my small part in this volume, I see things that I would like to change, things I would like to delete, but the political part that women have played in the fishing industry is an important part of so many things that are in effect today, and it will continue. Thanks to women, there is a new interest in the industry in information. In former days, the men who fished with one kind of gear had an association, and their wives usually were allowed to have an auxiliary. We worked hard to get the men's organizations to work together; it was much easier to get the women's organizations to work together. The women who worked in the seafood offices rarely had the opportunity to meet and talk "shop" with other women who worked in other seafood offices, with the exception of the group of women who had a loosely knit group called "The Sea Gals." This group met for a dinner meeting two or three times a year, with a speaker, but the main purpose was getting to know the women in other "fish offices."

Early in 1983 a group of bright, young, energetic women called together anyone interested in meeting with other women working within the fishing industry. I only attended two meetings then, but I could see what these women wanted to do and I understood their drive to become better educated in the various aspects of the fishing industry, to meet more people in the industry, and to talk shop. Soon it was evident that they had outgrown their small beginning. By mid-July of 1983, even with summer vacations, ninety women showed up for the dinner meeting. Now an average meeting includes between 120 and 150 women each month.

Each month they have an outstanding program that attracts both men and women. Some of the topics have been: allocation, role of fishery management, principles of warm-water aquaculture, transportation problems of fresh seafood, quality control problems, marketing updates, marine-vessel insurance, the use of and need for instruments in the wheel housing, maritime law, and many, many more. As many years as I have been in the industry, each month I look forward to learning something new at these meetings.

The women who have joined have usually had one thing in common— they have all worked in some phase of the fishing industry, which was a male-dominated industry that provided women very limited opportunities to grow or develop in their jobs. Some joined just to learn more about the industry, others for moral support, and others to upgrade their jobs. One of the original founders of the organization is an assistant manager of a local bank. She said it took her a long while to be accepted by fishermen as a "lending officer." Another is an editor of a seafood business report who has said that the network has made her job easier. If she needs information, she can go through the roster and find a Women's Fisheries Network (WFN) member in that company, and she has an immediate contact. Whatever the reason they joined, the organization is on the move and so are the women in it.

Coming out of the segment of the industry that I did, trying to get people together to understand one another's problems and work together to find solutions, I felt very strongly that the name for this group should be Northwest Fisheries Network or Seattle Fisheries Network, and not Women's Fisheries Network. Well, these gals don't need the men's names, or even to have it open to men. They now have a West Coast Chapter, Northeast Chapter, Northern California Chapter, and soon will have an Alaskan Chapter. They are doing great on their own, and my hat is off to all of them.

Not everyone is cut out to be a fisherman or fisherwoman, but everyone in the fishing industry can contribute something to keep this a viable industry, not only for us, but also for our children and our children's children. Fishing is not always lucrative, but it is a way of life, and that is how most people in the fishing industry view it. We chose this way of earning a living. We can get out of it if we want to, but most don't because they don't want to. There is an old joke that has been around for so long about the old fellow who was left a million dollars, and when he was asked what he planned to do with it, he just shrugged his shoulders and said, "Fish until I go broke, I guess." College students who spend their summers as crew members on fishing boats, even though the hours are long and the work is hard, will remember for the rest of their lives the clean air and the camaraderie of crew members, and they will always understand the independent feeling of being on the sea.

Fishing was America's first industry. We hope it will be again.

Appendix A
Consent Form

"Sociocultural Roles of Women in Commercial Fishing in the Puget
Sound Area"

Principal Investigators:
   Sue-Ellen Jacobs, Ph.D. Director and Ass't. Prof. Women
Studies (206) 543 6900
   Marc Miller, Ph.D. Ass't. Prof. Institute for Marine
Studies (206) 543 7004

Co-Investigators:
   Judith Hodgson, Director Women's Information Center
(206) 545 1090
   Karen J. Blair, Ph.D. Ass't. Prof. Women Studies
(206) 543 6984

Associates:
   Charlene Allison Women Studies (206) 543 6900
   Leona Pollock Women Studies (206) 543 6900

   Investigators' Statement

PURPOSES AND BENEFITS

We are part of a research team at the University of Washington. We are
conducting a study of women's roles in commercial fishing in the Puget
Sound area. We are interested in learning about and documenting for
historical purposes the many things women do as part of the commercial
fishing enterprise. We want to know, for example, whether or not women go
out on the boats with their husbands, fathers and/or sons during the
fishing season and what they do on the boats. When women stay home, what
do they do? Care for children? Manage the home and business matters?
Participate in fishermen's wives organizations; in what ways? There are
other questions along this line we want to ask.

   We believe that our study will provide very important insights into
the wide range of activities fisherwomen engage in in order to make fishing
a successful economic endeavor for their families. We also believe that
our study will result in publications that will encourage young women to
enter the commercial fishing industry.

PROCEDURES

   We are conducting interviews with people who are engaged in or who
once were involved in commercial fishing in the Puget Sound Area. I would
like to ask you some specific questions, including those listed above. I
would like to meet with you for one or two hours on a periodic basis over
the next three months to learn about your roles in commercial fishing. If
any of my questions seem rude or out of order, please tell me. You need
not answer any question which you feel is objectionable.

   Some interviews might be good to tape record, if you agree. After our
interviews have been completed (and at least within the next six months) I
will let you look over my notes so you can see if I have put everything
down right. You will be given copies of what is written down.

RISK, STRESS, OR DISCOMFORT

In some of the interviews I will ask questions about your family, your childhood experiences in fishing, education, economics, work and organizational involvements. You may choose not to answer any question. If my questions embarrass you or make you feel uncomfortable, please tell me. I do not want to make you uncomfortable in any way.

OTHER INFORMATION

Because the history of women in fishing in the Northwest has been neglected, we plan to develop an archival record of our research. We want to include everyone who is willing to be included in this archive. The archive will be a permanent record of women's involvement in fishing and be kept in the newly developed Washington Women's Heritage Project office at the University of Washington -- to be available to researchers for study. Tapes, photographs and all written records will be locked up and researchers will have conditional access to these.

You may choose to be listed, by name, and clearly identified in the archives. Or you may choose to remain anonymous; if you wish to NOT have your name listed, a code number will be used so that no one will know that you, as a named person, participated in our study. If you choose to have your name listed and/or to have photographs of you and your family available in the archives, you will be asked to sign a release form so that materials associated with you can be included in the permanent files of the Washington Women's Heritage Project.

You are free to withdraw from the study at any time without penalty.

```
        ------------------------------------------
        Signature of Investigator      Date
```

INTERVIEWEE'S STATEMENT

The study described above has been explained to me and I voluntarily consent to participate in this activity. I have had an opportunity to ask questions and agree to participate in the study. I agree___, do not agree___ that interview materials collected may be permanently located in the archives of the University of Washington (Women Studies) Washington Women's Heritage Project. I do___, do not___ wish to have my name listed.

```
        -------------------------------------
        Interviewee's signature      Date
```

Copies to: Interviewee
           Investigators' file.

Appendix B

Topics for Interviews
with Women in Commercial Fishing

Age

Early life
    Family/ where from
    Role of women in fishing families
    Specific women in family
    Relationships with siblings
    Women's organizations--family involvement
    Early images of fishing
    Religion
    Ethnic group

Education

Relationship with husband
    How met
    Marriage

Friendships with women

Fishing as a way of life: demands, special circumstances,
husband's time away

Organizational involvement:
    Women joined____: how org. is structured.
    Political evolution--education/who learned from, response
        of husband
    Changes in family life
    Changes in self-image

Sharing with other women/support systems

Role of women in political context--what is it?

Future of commercial fishing/women's role

Value systems/Life expectations

Child rearing
    Do they want their children to become fishermen?
    Problems of rearing children alone

Do the younger generation of women who fish want to have families?

Women's work
    Do men appreciate the work of women on the boat or at home?
    How do women value the work that they do - is it low status
    in comparison with men's work?
    What are the jobs that women perform?

Who makes decisions on a day to day basis?

Who does the family book-keeping/finances?

How do they view other fishermen?

Is crewing a transient occupation?/ Is there conflict between
crew members and the captain?

177

Appendix C
Questionnaire Used at Fish Expo

R:___M ___F

Interviewer_____

    I'm from the University of Washington. We are doing a study on the activities of women in commercial fishing and would like to ask you a few questions. First of all,

1. How are you involved in commercial fishing?

2. Which fisheries (geartype/species/location) are you primarily interested in?

3. What city do you live in?

4. I'm going to read a number of categories. Please tell me if you know of any women who are involved in commercial fishing in that capacity (where have you seen women in this capacity - in terms of port, fishery and/or gear type?)

Yes   No   a)Women who own commercial fishing boats either solely __, or as partners __. (Record which)
                  Where?

Yes   No   b)Women who captain commercial fishing boats.
                  Where?

Yes   No   c)Women who handle gear on commercial fishing boats.
                  Where?

Yes   No   d)Women who cook food on commercial fishing boats.
                  Where?

Yes   No   e)Women who work on tenders.
                  Where?

Yes   No   f)Women who are on boats solely as wives or friends and take no part in fishing or cooking.
                  Where?

Yes   No   g)Women who cook at fishing camps.
                  Where?

Yes   No   h)Women who captain sports fishing boats.
           Where?

Yes   No   i)Women who crew on sports fishing boats.

                  Where?

Yes   No   j)Women who are managers in the fish processing
              industry.
              Where?

Yes   No   k)Women who work lines in canneries and cold storage.
              Where?

Yes   No   l)Women who work as salespeople in the wholesale
              marketing sector.
              Where?

Yes   No   m)Women who work in advertizing in the marketing
              sector.
              Where?

Yes   No   n)Women who make fishing equipment, for example
              crab pots.
              Where?

Yes   No   o)Women who are net menders/hangers.
              Where?

Yes   No   p)Women who are salespeople in gear stores.
              Where?

Yes   No   q)Women who work on gas docks.
              Where?

Yes   No   r)Women who are in fishery regulations enforcement.
              Where?

Yes   No   s)Women who are marine (ship-shore) operators.
              Where?

Yes   No   t)Women in fishery research.
              Where?

Yes   No   u)Women who are politically active in furthering the
              interests of commercial fishermen. (e.g.)_____
              Where?

Yes   No   v)Women who work in fund raising, for example selling
              cookbooks.
              Where?

Yes   No   w)Women who are novelists or historians reflecting the
              fishing industry.
              Where?

5. Are there any other ways in which women are involved in or
promoting the fishing industry that I have not mentioned?

6. What is your age?

7. Would you be willing to help us further in later parts of this
study? ___Yes ___No.

8. Could we have your name and phone number so that we can
contact you in the future?

Appendix D

Place-Name Locater

This listing includes places referred to in the life histories. Longitude and latitude are from *The Great Geographical Atlas* (Chicago: Rand McNally, 1984), except those marked*, which were calculated by project personnel, and those marked **, which are from *The Times Atlas of the World, Comprehensive Edition* (New York: Times Books, 1983).

Aberdeen, WA	46.59N 123.50W
Adak Island, AK	51.45N 176.40W
Afognak Island, AK	58.15N 152.30W
Akutan, AK	54.08N 165.46W
Alaska	State
Alaska, Gulf of	58.00N 146.00W
Aleutian Islands, AK	52.00N 176.00W
Alitak, AK	56.51N 154.21W
Amaknak Island, AK	53.53N 166.32W
Anchorage, AK	61.13N 149.53W
Astoria, OR	46.11N 123.50W
Attu Island, AK	52.56N 174.45E
Bainbridge Island, WA	47.42N 122.55W*
Bellingham, WA	48.49N 122.29W
Bering Sea	West of Alaska
Bristol Bay, AK	58.00N 159.00W
California	State
Camano Island, WA	48.19N 122.30W*
Cape Cook, BC	50.10N 127.54W*
Cape Fairweather, AK	58.50N 137.55W*
Cape Flattery, WA	48.23N 124.46W*
Cape Spencer, AK	58.14N 136.57W*
Charleston, OR	43.24N 124.24W*
Chignik Lagoon, AK	56.14N 158.44W
Columbia River, OR/WA	46.45N 119.05W
Cook Inlet, AK	61.00N 151.00W*
Cordova, AK	60.33N 145.46W
Craig, AK	55.29N 133.09W
Cross Sound, AK	58.10N 136.30W
Dry Spruce Island, AK	57.58N 153.00W*
Dutch Harbor, AK	53.53N 166.32W
Duwamish Waterway, WA	47.34N 122.21W*
Everett, WA	47.59N 122.14W**
Fairbanks, AK	64.51N 147.43W

Fairweather Grounds, AK	58.30N 139.00W*
False Bay, WA	48.28N 123.04W
Farallon Islands, CA	37.43N 123.03W
Fife, WA	47.16N 122.20W*
Forks, WA	47.57N 124.23W*
Fortuna, CA	40.36N 124.09W
Fraser River, BC	49.15N 123.00W*
Friday Harbor, WA	48.32N 123.01W
Georgia Strait, WA/BC	48.70N 123.01W*
Grays Harbor, WA	46.56N 124.05W
Hoh River, WA	47.45N 124.29W
Homer, AK	59.39N 151.33W
Hood Canal, WA	47.35N 123.00W
Hoquiam, WA	46.56N 123.53W
Humboldt Bar, CA	40.47N 124.11W
Ilwaco, WA	46.19N 124.02W*
Inside Passage, BC	Waters between Vancouver Island and the Canadian mainland.
James Island, WA	47.54N 124.39W*
Kenai Peninsula, AK	60.10N 150.00W
Ketchikan, AK	55.21N 131.35W
Kiska Island, AK	52.00N 177.30E
Kodiak, AK	57.48N 152.23W
Koitlah Point, WA	48.23N 124.39W*
Kukak Bay, AK	58.20N 155.15W*
Lake Washington, WA	47.40N 122.15W*
La Push, WA	47.54N 124.38W*
Little Boston (Port Gamble, WA)	47.51N 122.35W
Lopez Island, WA	48.30N 122.50W*
Marysville, WA	48.03N 122.11W
Moss Landing, CA	36.50N 121.46W*
Neah Bay, WA	48.22N 124.37W
Newport, OR	44.38N 124.03W
Ocean Shores, WA	46.58N 124.09W*
Olympic Mountains, WA	47.50N 123.45W
Olympic Peninsula, WA	47.50N 123.45W
Oregon	State
Pelican, AK	57.57N 136.13W
Petersburg, AK	56.50N 132.59W
Port Angeles, WA	48.06N 123.26W**
Port Gamble, WA	47.51N 122.35W
Portland, OR	45.33N 122.36W
Port Lions, AK	57.52N 152.53W
Port Townsend, WA	48.09N 122.48W

Pribilof Islands, AK	57.00N 170.00W
Prince Rupert, BC	54.19N 130.19W
Puget Sound, WA	47.50N 122.30W
Queen Charlotte Sound, BC	51.30N 129.30W
Richmond Beach, WA	47.46N 122.22W
Russian River, CA	38.27N 123.08W
San Francisco, CA	37.48N 122.24W
San Juan Island, WA	48.28N 123.00W
San Juan Islands (Archipelago), WA	48.36N 122.50W
Seattle, WA	47.36N 122.20W
Sekiu River, WA	48.17N 124.24W*
Seldovia, AK	59.27N 151.43W
Sequim, WA	48.05N 123.06W
Seward, AK	60.06N 149.26W
Shelikof Straits, AK	57.30N 155.00W*
Sitka, AK	57.03N 135.40W
Skagway, AK	59.28N 135.19W
Spokane, WA	47.40N 117.23W
Strait of Juan de Fuca, WA/BC	48.18N 124.00W
Swiftsure Bank, BC	48.32N 125.00W*
Tacoma, WA	47.15N 122.27W
Tatoosh Island, WA	48.23N 124.44W*
Togiak, AK	59.04N 160.24W
Umatilla Reef, WA	48.12N 124.47W*
Unalaska, AK	53.52N 166.32W
Vancouver Island, BC	49.45N 126.00W
Washington	State
Westport, WA	46.53N 124.06W
Whidbey Island, WA	48.15N 122.40W
Wishkah, WA	47.01N 123.50W*
Yakutat, AK	59.29N 139.49W**

Appendix E
Map

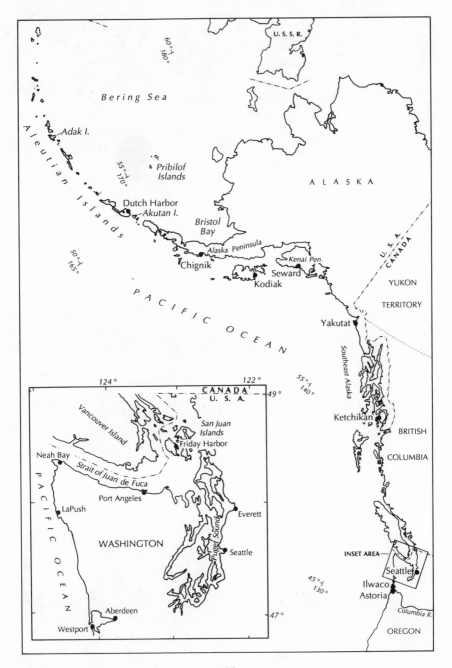

Glossary

The definitions in this glossary have been formulated primarily to explain the terms as they are used in the life histories. Other, equally correct definitions may have been omitted.

Alaska Trollers Association: founded in 1925. An organization of Alaskan power and hand troll permit owners dedicated to maintaining the economic viability of Alaskan trolling through preserving salmon spawning areas, monitoring legislation, and participating in ongoing U.S.-Canada treaty negotiations.

albacore: *Thunnus alalunga*, a tuna with long pectoral fins. Also called longfin.

Aleut people: original inhabitants of the Aleutian islands and parts of the Alaskan mainland. Aleut people have occupied the Aleutian Islands continuously for 9,000 years.

anadromous: describes fish that migrate from fresh to salt water and back to fresh water for breeding.

anchor winch: the horizontal drum or cylinder around which the cable holding the anchor is wound, and by which the anchor is raised and lowered.

baching: housekeeping by an unmarried man.

banana sets: in beach seining, the long, narrow sets formed by taking the net straight out from the shore or boat, then bringing the net's end straight back in.

bar: a bank of sand or gravel, often found at the mouth of a river, which may be in shallow water or partially above the water.

beach seine boats: boats used for beach seining.

beach seining: a form of drag seining. A small boat goes out from the shore and circles back in, letting out net as it goes. The circle of net is then dragged in to the shore, pulling in the trapped fish.

belly burn: deterioration in quality of fish if they are not cleaned and/or adequately iced soon after being caught.

blackmouth: immature chinook salmon.

blind set: setting up a net on a hunch, when there is no physical indication of fish (such as gulls gathering or fish jumping).

blue crab: *Paralithodes platypus*, one of three subspecies of king crab.

boatpuller: deckhand, on troller or gillnetter, whose main responsibility is bringing in the fish.

Boldt case area: *United States v. Washington* case area. The area covered by the ruling issued as a result of the Boldt Decision.

Boldt Decision: *United States v. Washington*, 1974. A U.S. District Court ruling, which provides that the American Indian tribes party to a series of treaties signed in 1854 and 1855 are entitled to an equal share of the catch and to self-management of fisheries in traditional fishing grounds.

bonito: saltwater fish of the mackerel family, related to the tuna.

bottomfish: any of a number of fish that live near the ocean floor, (e.g., sole, tur-
bot, flounder).

brail: bringing fish on board using a brailer.

brailer: a large dip net or basket used to scoop fish on board ship from a tender or
a pursed seine net.

broker: a person hired to make contracts and sales on behalf of a company.

bulkhead: an inside wall constructed to strengthen a boat or divide it into com-
partments.

bull wincher: in beach seining, operator of a large winch located on shore, through
which the net cables are drawn and let out.

buoy system: an arrangement of floating markers of different shapes, colors, lights,
and sounds, which aid navigation by marking channels and warning of obstacles.

butcher bin: site where crab or fish are butchered before processing.

candy striper: teen-age girl who does volunteer hospital work.

cannery: a factory where fish are canned. The term is also used colloquially to refer
to any fish-processing establishment, even where fish are frozen rather than canned.

capstan: a vertical winch.

CB: citizen's band radio. Also called a *Mickey Mouse.*

chalk off a line: a technique used in construction to ensure a straight line—a string
or rope is rubbed with chalk and then sharply pulled taut, leaving a straight chalk
line.

Charter Boat Association: see Washington State Commercial Passenger Fishing Vessel
Association.

chinook: *Oncorhynchus tschawytscha,* the largest species of Pacific salmon. Average
weight twenty pounds. Mature at two to eight years. Also called *king* or *tyee* salmon.
Chinook that spawn in the spring as opposed to the fall are called *spring* salmon.
See also *blackmouth.*

chum: *Oncorhynchus keta,* a species of Pacific salmon. Average weight ten to twelve
pounds. Mature at three to five years. Usually caught by netfishing rather than
trolling. Also called dog or fall salmon. Chum is also a term for live bait fish.

clam: an edible, bivalve, hardshelled mollusk.

clipping clams: preparing shelled clams for canning by clipping off the end of the
syphon and slitting the body from the base of the foot to the end of the syphon to
remove the viscera.

code group: a group of fishermen on separate boats who share information about
fishing conditions over the radio, using code words so other fishermen will not know
their meanings.

coho: *Oncorhynchus kisutch,* a small Pacific salmon. Average weight eight to twelve
pounds. Mature at two to three years. Trolled coho are sold fresh or frozen; gillnetted and seined coho are usually canned.

convoy: a group of boats traveling together for mutual protection.

cooker: a very large container full of boiling water which cooks thousands of pounds
of crab at one time.

corking: setting your own net between the fish and another fisherman's net.

corkline: a line with floats distributed along it that keeps a fishing net suspended in
the water.

corks: floats used on a corkline.

custom work: packing and processing done by someone other than the owner of the fish; also work done to customer specifications.

dinghy: a small rowboat or sailboat.

dogfish: any of various small sharks, especially those in the squalidae family.

double-ender: a boat, usually a troller or gillnetter, with a bow and stern of the same shape; both ends are pointed.

dragger: a boat that catches a variety of fish by pulling a net behind the boat, close to the sea bed, for several hours at a time.

dredge: equipment for moving mud, sand, or rock in order to deepen or clear channels and harbors.

drifting: a boat being carried with the current.

drift net: a net used for drift-netting, which is a form of gillnetting.

fathom: a unit of length equal to six feet, usually used to measure water depth.

Fish Expo: annual exhibition of the latest fishing equipment held alternately in Seattle and Boston, with workshops on fishing issues and techniques.

fishing-processor: a ship that both catches and processes fish.

fish traps: structures built in the water to entrap fish, often built with wood piles and some kind of lattice work (e.g., wire mesh). Now illegal in both Alaska and Washington.

fleet: term used to refer to any group of boats (e.g., the Pacific Coast fleet, the gillnet fleet); also, a group of fishing friends in a code group. See code group.

float bags: devices on trollers which float wires and weights behind the boat to help keep the lines separated.

flume: an inclined channel or water-chute, used in fish-processing for conveying fish.

fo'c'sle: the forecastle or forward section of a boat, where the living quarters are often located.

freshet: a rush of fresh water flowing into the sea, sometimes from springtime snowmelt.

fry: young fish.

Gear Buy-back Program: Washington Department of Fisheries program which buys back salmon fishing licenses from nontreaty fishermen to reduce the size of the nontreaty salmon-fishing fleet.

gillnetter: a boat used for gillnetting; also, a person who fishes with a gillnet.

gillnetting: a form of net-fishing. The net is held vertically in the water by a corkline along the top and a leadline along the bottom. Fish become caught by their gills when they try to swim through the net. The net is made of fine nylon, so it is not visible to the fish. The holes are large enough that fry can pass through. There are two types of gillnetting—drift gillnetting and set-netting.

glaze: a fine ice coating put on a fish, after it is frozen, to prevent dehydration. The glaze contains a sugar solution (e.g., fructose or corn syrup).

grading: sorting fish according to quality, which ranges from perfect (no bones sticking out and no knife marks or bruising on the skin) to sour (very soft and bruised and with belly burn).

Grange, the: the National Grange of the Patrons of Husbandry, an organization founded in 1867 by American farmers to promote their interests.

greenbelt: a moving conveyor belt at which lineworkers, usually women, stand to clean or pack fish.

Gresen valve: a hydraulic valve manufactured by the Gresen Company.

gunnel: the gunwale, the upper edge of a ship's or boat's side.

gurdies: large reels on a troller which let out, and pull in, the fishing lines. They are usually power operated.

hake: bottomfishes of *Merlucciidae* family.

halibut: *Hippoglossus stenolepis,* a large flatfish found in northern Pacific seas, sometimes weighing hundreds of pounds.

Halibut Producers Cooperative: founded in 1944, a cooperative of fishermen who own their own processing plant and sell their catches to it.

handline: trolling using a hand-powered gurdy rather than electric power.

hanging nets: assembling seine nets or gillnets. A net is laid out along the ground, and a section of it is hung on a post. That section is then assembled with the cork- and leadlines and other attachments. When that section is assembled, it is moved to the ground, and a new section is hung.

herring: bony fishes in *Clupeidae* family.

highliner: a fisherman who, through luck and skill, consistently catches large quantities of fish.

high slack: high-water slack. A period, associated with high tide, when there is no current in the water. The point between flood, when the tide moves in, and ebb, when the tide moves out.

hooknose silver: coho salmon.

humpie: pink or humpback salmon.

ice hold: storage space in a boat where the fish are stored in ice.

icing down: putting fish in the hold with layers of ice in order to keep them fresh.

International Pacific Salmon Fisheries Commission: a joint United States–Canada commission established in 1937 to manage and enhance the sockeye and pink salmon fisheries in the Fraser River.

jig: a feathered lure used to catch salmon or tuna.

jigging: fishing by trailing jigs through the water or bouncing them along the bottom.

Jimmy-come-lately: a colloquialism for newcomer.

kelper: a small boat used for fishing in close to shore; also, the operator of that boat.

ketch: a sailing vessel with a mainmast toward the bow and a relatively tall mizzenmast forward on the rudderpost, near the stern.

king crab: *Paralithodes camtschatica,* a large species of northern Pacific crab. Average weight eight pounds. Mature at eight years. Only males can be kept.

king crab season: the fall months. Exact dates are decided by regulatory agencies.

king salmon: chinook salmon.

knot: a unit of speed of one nautical mile (6,076.10 feet) per hour.

leadline: a marked line with a lead weight at one end, used to measure water depth; also, a lead-weighted line along the bottom edge of a fishing net.

Liberty ship: a type of U.S. general cargo ship used during World War II to transport troops and supplies. Some of the ships were converted to fish-processing factories after the war.

limited entry: a program in Alaska to reduce the number of commercial salmon fishermen by issuing a limited number of entry permits.

lingcod: *Ophiodon elongatus,* closely related to the greenling family, a greenish-fleshed bottomfish found as far south as Baja California and as far north as Alaska.

longfin: see albacore.

Loran: long-range navigation, an electronic navigational system by which users pinpoint their precise location on a chart.

lure: a floating device, often metal, used to attract fish.

mackinaw: a heavily napped and felted woolen cloth coat or blanket, usually plaid.

marine gillnet boat: a gillnet boat that fishes in salt water rather than rivers.

mean tide: average water level, halfway between low and high tide.

Mickey Mouse: see CB.

midships: amidships, in the middle of the boat.

minesweeper: a ship that seeks and immobilizes mines at sea.

mooch: to troll for fish close in to shore with sportfishing equipment, including a *spoon.*

National Federation of Fishermen: founded in 1969 through the merger of two West Coast organizations, the Congress of American Fishermen and the National Fishermen and Wives. It is a national federation of commercial fishermen's organizations with representation in Washington, D.C., since 1973.

National Fishermen and Wives: founded in 1963 by the women's auxiliaries of the gillnet, halibut, and trolling organizations. Merged in 1969 to become National Federation of Fishermen.

necking down: Attaching a larger pipe to a smaller one using a transitional size (or sizes) of pipe as a connector.

nontreaty fishermen: any fishermen, whether Indian or non-Indian, who are not members of those Indian tribes party to the treaties upheld by the Boldt Decision.

Northwest Indian Fisheries Commission: formed in 1974 as a result of the Boldt Decision, an alliance of the western Washington treaty tribes to organize effective management of their fisheries resources and to promote public awareness of treaty fishing rights and tribal fisheries management.

opilio: *Chionoecetes opilio,* one of three kinds of crab known collectively as tanner crab.

outside coast: the western coast of Vancouver Island, as opposed to the Inside Passage.

Pacific Coast Fishermen's Wives Coalition: founded in 1976, a coalition of women's fishing organizations to promote the interests of commercial fisheries.

Pacific Troller Association: founded in 1968 (formerly Washington Kelpers Association). Promotes resource enhancement and the interests of trollers and other small (under thirty feet) boat operators.

pack: the total production of fish for one year.

pan up: to put fish in large stainless steel pans before freezing them.

pewing: using a *pew* (a long wooden-handled tool with a curved spike on the end) to unload fish from a fishing boat.

pick fish: to remove fish from a gillnet.

pile driver: a machine with a drop hammer for driving piles.

pinks: *Oncorhynchus gorbuscha,* a small pink salmon. Average weight five to six pounds when mature. It has a life cycle of only two years. Also called humpbacks or humpies.

pit: the cockpit on a troller where a fisherman operates the fishing gear.

plug: a plastic lure supposed to resemble a herring.

Point Elliott Treaty Area: an area covered by the 1855 Treaty of Point Elliott, made between the United States and the Lummi, Muckleshoot, Nooksack, Sauk-Suiattle, Stillaguamish, Suquamish, Swinomish, Tulalip, and Upper Skagit tribes. It includes approximately the waters of the western portion of the Puget Sound area from the northern edge of Mount Rainier (including most of the White River, the Green River, but not including the area around the city of Tacoma) to the south and extends to the Canadian border on the north and northwest.

Point Elliott Treaty Council: a council of representatives from the tribes party to the 1855 Treaty of Point Elliott, who send a representative to the Northwest Indian Fisheries Commission.

poles: a pair of long poles extending out from the sides of a troller.

pots: steel king-crab pots, weighing six to eight hundred pounds. Crabs are attracted into the pots by fish bait and then cannot escape.

princess dressing: removing gills and guts from a salmon.

processing: any procedure done to fish to prevent its deterioration before reaching the consumer (e.g., freezing, canning, drying).

processor: a ship on which fish are processed.

prop: propeller

Puget Sound Gillnetters Association: founded in 1948 to promote the interests of Puget Sound gillnetters in fisheries regulation and legislation.

Puget Sound Gillnetters Association Women's Auxiliary: founded in 1954, women's organization that raises funds and promotes the political and commercial interests of Puget Sound gillnetters.

Puretic block: a hydraulic device for lifting seine nets out of water.

purser: the person on board ship who is in charge of accounts, freight, etc.

purse seiners: boats, or people, that go purse seining.

purse seining: fishing with a very long seine net, around 1,500 feet. The net is pulled in a circle by a skiff, thus surrounding the fish. A purse line runs underneath the net. When it is pulled up, the net is gathered in like a closing purse and the fish are trapped. The net is then brought on board to unload the fish, or, if the load is too heavy, the fish are brailed on board.

PX: the post exchange, a store selling goods and services to armed forces personnel and to others with authorization.

Quileutes: American Indians who live at La Push at the mouth of the Quillayute River.

raw tanner pack: production of tanner crab that is uncooked.

reds: sockeye salmon.

round weight: the weight of fish before it has been cleaned.

rigging: the chains, ropes, etc., that support the mast and sails on a boat.

run: fish from a common stock traveling together to breed in a particular river or lake.

running off the deck: the job done by a purse-seine crew member who stays on deck, while the skiff people take the net out, to ensure that the net unrolls smoothly.

salmon: the Pacific salmon (family *Salmonidae*, genus *oncorhynchus*), with silver scales and flesh that is yellowish-pink to pale red when cooked. Salmon hatch in fresh water, live in salt water, and return to spawn in fresh water at the end of the life cycle.

scaling: removing scales from fish.

scallop: *Pectinidae pecten*, a marine bivalve mollusk with grooved shells. It swims by rapidly snapping its shells together.

scupper: a drainage hole in a boat's side that allows water to run off the deck.

seine: a portable fishing net used to encircle fish, also, to fish with a seine net.

seiners: fishermen who fish with seines.

seining: to encircle fish with a seine net. See beach seining, purse seining.

set: to place a net in position to catch fish. To make a set is to make such a placement.

set-netting: type of fishing practiced in Alaska and by American Indians in Washington State. One end of the fishing net is fixed to the shore. The other end is anchored out in the water. Fish are caught by their gills as they swim through the net, which is held vertical by a leadline along the bottom and a floating corkline along the top.

shrimp: a small marine crustacean.

side trawler: a boat, of any size, that sets and recovers a bag-shaped net (a trawl) over the side of the vessel.

silvers: coho salmon.

skiff: a small boat used in seining to position the net in the water.

skiff people: crew who operate the skiff in seining.

sliming: cleaning all the guts out of a fish, including the fine membrane attached to the meat.

snow crab: marketing name for tanner crab.

sockeye: *Oncorhynchus nerka*, a Pacific salmon. Average weight five to seven pounds, Mature at three to six years. Normally caught by net fisheries and canned. Also called *red salmon*.

Southeastern: an area of Alaska covering the coast and islands of the Alexander Archipelago, from the United States–Canadian border north to Cape Fairweather; also, Southeast.

Sou'wester: a waterproof hat with a wide brim at the back to protect the neck.

splash boy, splash girl: young person in Tulalip beach seining who splashes the water and thus forces the fish to go in the direction of the fishing nets.

spoon: a fishing lure, usually made of metal, which lies above the hook and wobbles when drawn through the water.

steelhead: an anadromous rainbow trout found along the Pacific Coast. Spawns in rivers. It is caught by American Indians and sportfishermen only.

straight tack: a straight course.

stock: fish that breed together in a particular river or lake at the same time.

stumpage money: income from felled timber.

tanner crab: name used to refer to three kinds of crab: *Chionoecetes opilio*, *Chionoecetes bairdi*, and *Chionoecetes tanneri*. *Opilio* and *bairdi* are caught in Alaska, *tan-*

neri is found farther south in Canadian waters. Tanner is smaller than king crab, weighing about five pounds, and was not commercially exploited until the mid-1960s.

tanner crew: personnel who process tanner crab.

tender: a boat that services fishing boats and processors by transporting fish from the fishing fleets at sea to floating or land-based processors.

tie off: to cleat or secure a line.

trawl: a bag-shaped net used in trawling.

trawling: towing a net through the water on the sea bottom or close to it to entrap fish. Also called dragging. There are two methods of trawling—side trawling and stern trawling.

treaty fishermen: Indian fishermen whose tribes are party to the treaties upheld by the Boldt Decision.

trip-fishing: fishermen staying out at sea fishing for up to ten days before returning to port.

troller: a boat that catches fish by trolling; also, a person who trolls.

Trollers Association: Pacific Trollers Association or Washington Trollers Association.

trolling: fishing with hooks and lines trailed through the water behind a boat.

trout: Member of salmon family, but smaller than related salmon; found mainly in fresh water.

Tulalip: American Indian tribes now living on the Tulalip Indian Reservation northwest of Marysville, Washington.

tuna: *Thunnus thynnus*, a large ocean fish of the mackerel family, with oily flesh; also, other species of tuna, including albacore.

twelve-mile limit: an exclusive fishing zone extending twelve miles from the U.S. coast, set up in 1966 by adding a nine-mile zone to the United States's already existing three-mile territorial sea.

two hundred-mile limit: two hundred-mile Fisheries Conservation Zone established by the Magnuson Fisheries Conservation Management Act of 1976. Foreign fleets cannot fish within the zone, except by permit.

tyee: chinook salmon.

USO: United Service Organizations: An independent, voluntary organization founded in 1941 to meet recreational, spiritual, and social welfare needs of U.S. service personnel.

WAAC: Women's Auxiliary Army Corps. Founded in 1942, a forerunner of the Women's Army Corps.

WAC: Women's Army Corps, founded in 1943.

Washington Association for Fisheries: founded in 1971 (formerly Seattle Association for Fisheries) to promote the upgrading of fisheries management and maintain the fisheries as a viable resource in Washington State.

Washington State Commercial Passenger Fishing Vessel Association: founded in 1974 to promote the interests and continued economic viability of ocean sportfishing boat operators. Also referred to as Washington State Charter Boat Association. It is a coalition of Westport, Ilwaco, and Olympic Peninsula Charter Boat Associations.

Washington Trollers Association: founded in 1978 (formerly West Coast Trollers

Association). Political action organization composed of fishermen operating larger trolling vessels in Alaska, California, Oregon, and Washington. Promotes the trolling industry on the West Coast.

welfare: used colloquially to refer to any governmental program providing aid to the poor or disabled.

wheelhouse: an enclosed place on the upper deck of a boat or ship, where the helmsman stands to steer.

whistle punk: a starting job in logging, traditionally done by a boy. The whistle punk passes signals, by hand or by whistle, between the logger who attaches cables to the logs and the logger who runs the donkey engine, which provides the power for dragging logs into the logging yard from the stumps.

winch: a vertical or horizontal spindle-mounted drum that moves heavy weights by drawing in or letting out a cable.

Yakima: Sahapatian-speaking American Indian tribes now living on the Yakima Indian Reservation in south central Washington.

Bibliography

Adasik, Allan
  1978 "The Alaskan Experience with Limited Entry." In *Limited Entry as a Fisheries Management Tool*, eds. R. Bruce Rettig and Jay J. C. Ginter, pp. 271–99. Seattle: Washington Sea Grant.

Alaska Fisherman
  1979 "Fish Expo Attracted 16,000," vol. 7, no. 3 (November): 21.

Alaska, state of, Governor William A. Egan
  1973 "A Limited Entry Program for Alaska Fisheries." *Report of the Governor's Study Group on Limited Entry.* Juneau.

Allison, Charlene J.
  1988 "Women Fishermen in the Pacific Northwest." In *To Work and to Weep: Women in Fishing Economies*, eds. Jane Nadel-Klein and Dona Lee Davis. St. John's: Institute for Social and Economic Research, Memorial University of Newfoundland.

Alverson, D. L., and L. D. Lusz
  1967 *Electronics Role in Fishing Industry—Present and Future.* IEEE International Convention Record: Part 8, Instrumentation Ocean Electronics Highlight Symposium. New York: The Institute of Electrical and Electronics Engineers.

Alverson, D. L., and A. T. Pruter
  1980 "Nature's Impact on Fisheries Management." *The Fishermen's News*, vol. 36, no. 13 (June):8–9, 21, 28.

American Friends Service Committee
  1970 *Uncommon Controversy.* Seattle: University of Washington Press. Reprinted with new preface, 1975.

Anderson, Myrdene
  1977 "Woman as Generalist, as Specialist, and as Diversifier in Saami Subsistence Activities." *Humboldt Journal of Social Relations* 10:175–97.

Antler, Ellen
  1977 "Women's Work in Newfoundland Fishing Families." *Atlantis* 11:106–13.

Armitage, Susan
  1982 "Restoring Women to Pacific Northwest History." In *A Handbook for Life History Research*, eds. Sue-Ellen Jacobs, Susan Armitage, and Kathryn Anderson. Seattle: Washington Women's Heritage Project at the University of Washington.

Bagley, Clarence B.
  1916 *History of Seattle.* Vol. 1. Chicago: The S. J. Clarke Publishing Company.

Barsh, Russel L.
  1979 *The Washington Fishing Rights Controversy: An Economic Critique.* Seattle: University of Washington, Graduate School of Business Administration.

Bartlett, Craig
  1980   "A Talk With Jim Heckman of the NW Indian Fishing Commission." *The Fishermen's News*, vol. 36, no. 13 (June):16–17, 25.
  1981   "Phase II: A Weapon for Fisheries?" *The Fishermen's News*, vol. 37, no. 4 (February): 14–16.
Bell, Douglas
  1978   "Gear Reduction/Buyback Programs in British Columbia and Washington State." In *Limited Entry as a Fisheries Management Tool*, eds. R. Bruce Rettig and Jay J. C. Ginter, pp. 353–57. Seattle: Washington Sea Grant.
Benson, Gary, and Robert Longman
  1978   "The Washington Experience with Limited Entry." In *Limited Entry as a Fisheries Management Tool*, eds. R. Bruce Rettig and Jay J. C. Ginter, pp. 333–52. Seattle: Washington Sea Grant.
Blair, Karen
  1980   *The Clubwoman as Feminist.* New York: Holmes and Meier.
Boeri, David
  1980   "Boldt Decision May Dog WA Trollers in Alaska." *Alaska Fisherman's Journal*, vol. 3, no. 4 (April):1–22.
Boldt, Judge George H.
  1974   *United States v. Washington* (Phase I). U.S. District Court, Western District of Washington at Tacoma. Civil no. 9213. *Federal Reporter*, 384 F. Supp. 312.
Brief of Respondent Indian Tribes in the Supreme Court of the United States
  1978   *State of Washington v. Washington State Commercial Passenger Fishing Vessel Association and Washington Kelpers Association.* October Term.
Browning, Robert J.
  1974   *Fisheries of the North Pacific: History, Species, Gear and Processes.* Anchorage: Alaska Northwest Publishing Company.
Campo, Joe
  1983   *1979 Summary of Commercial Salmon Fishing Regulations in Puget Sound.* Olympia: Washington Department of Fisheries, Harvest Management Division.
Carter, Glen
  1982   "Salmon, Crab Aren't Lucrative Anymore." *Seattle Times*, January 10:C3
Christensen, James B.
  1977   "Motor Power and Woman Power: Technological and Economic Change Among the Fanti Fishermen of Ghana." In *Those Who Live From the Sea: A Study in Maritime Anthropology*, ed. M. Estelle Smith, pp. 71–96. American Ethnological Society Monograph 62. St. Paul: West Publishing Company.
Cobb, John
  1911   *The Salmon Fisheries of the Pacific Coast.* Bureau of Fisheries Document No. 751. Department of Commerce and Labor, Bureau of Fisheries. Washington, D.C.: U.S. Government Printing Office.
Cohen, Fay G.
  1986   *Treaties on Trial.* Seattle: University of Washington Press.

Connell, J. J., ed.
  1980 *Advances in Fish Science and Technology.* Farnham, Surrey, England: Fishing
        News Books, Ltd.
Cook, Alice Hanson
  1984 "Women and Work in Industrial Societies." In *Urbanism and Urbanization,*
        ed. Noel Iverson, pp. 194–224. The Netherlands: E. J. Brill.
Cooley, Richard A.
  1963 *Politics and Conservation: The Decline of the Alaska Salmon.* New York: Har-
        per and Row.
Crutchfield, James A., and Giulio Pontecorvo
  1969 *The Pacific Salmon Fisheries: A Study of Irrational Conservation.* Baltimore:
        The Johns Hopkins Press for Resources for the Future.
Danowski, Fran
  1980 *Fishermen's Wives: Coping with an Extraordinary Occupation.* Marine Bulletin
        37. Kingston, Rhode Island: NOAA Office of Sea Grant, University of Rhode
        Island.
Droker, Howard A.
  1979 *Puget Sound Salmon Fishermen.* City of Seattle.
Dunlap, G. D.
  1972 *Navigating and Finding Fish with Electronics.* Camden, Maine: International
        Maine Publishing Co.
Ellis, Carolyn
  1984 "Community Organization and Family Structure in Two Fishing Commu-
        nities." Manuscript. Earlier draft read at the Southern Sociological Society
        meeting, April 1983.
Fishermen's News, The
  1980 "Proposed 1980 Ocean Salmon Fishing Regulations as Compared With 1979
        Regulations," vol. 36, no. 6 (March):2.
  1980 "Buy-Back Aid," vol. 36, no. 19 (October):31.
  1980 "Excerpts From the Judgement," vol. 36, no. 21 (November):33, 35.
  1980 "Buy-Back Offers a Way Out," vol. 36, no. 22 (November):30–31.
Geiger, Susan N. G.
  1986 "Women's Life Histories: Method and Content." *Signs: Journal of Women
        in Culture and Society* 11:334–51.
Gillies, M. T.
  1975 *Fish and Shellfish Processing.* Park Ridge, New Jersey: Noyes Data Corpora-
        tion.
Henry, Kenneth A.
  1981 "Pacific Salmon Interception." In *Science, Politics and Fishing: A Series of
        Lectures,* ed. Naomi Krant, pp. 137–50. Corvallis: Sea Grant College Pro-
        gram, Oregon State University.
Higgins, John
  1978 *The North Pacific Deckhand's and Alaska Cannery Worker's Handbook.* East-
        sound, Washington: Albacore Press.
Ives, Edward D.
  1980 *The Tape-Recorded Interview: A Manual for Fieldworkers in Folklore and Oral
        History.* Knoxville: University of Tennessee Press.

Jacobson, Jon L.
  1981 "International Law of the Sea and Other Legal Implications of Fisheries
        Management." In *Science, Politics and Fishing: A Series of Lectures*, ed. Na-
        omi Krant, pp. 11–22. Corvallis: Sea Grant College Program, Oregon State
        University.
Jurkovich, Jerry E.
  1981 "Principles and Innovations in Commercial Fishing Gear." In *Science, Pol-
        itics and Fishing: A Series of Lectures*, ed. Naomi Krant, pp. 23–32. Cor-
        vallis: Sea Grant College Program, Oregon State University.
Kanter, Rosabeth Moss
  1977 *Work and Family in the United States: A Critical Review and Agenda for Re-
        search and Policy*. New York: Russell Sage.
Langness, L. L., and Gelya Frank
  1981 *Lives: An Anthropological Approach to Biography*. Novato, California: Chan-
        dler and Sharp.
Margolis, Maxine
  1984 *Mothers and Such*. Berkeley: University of California Press.
Martin, Roy E., ed.
  1982 *Proceedings of the First National Conference on Seafood Packaging and Ship-
        ping*. Washington, D.C.: Science and Technology National Fish-
        eries Institute.
Matsen, Brad
  1981 "Troll Salmon Cut-back Dilemma: 'Not Enough Fish For All the Actors.' "
        Alaska Fisherman's Journal, vol. 4, no. 4 (April):1–18.
McHugh, J. L.
  1984 *Fishery Management*. Lecture Notes on Coastal and Estuarine Studies 10.
        New York: Springer-Verlag.
Meltzer, Michael
  1980 *The World of the Small Commercial Fishermen: Their Lives and Their Boats*.
        New York: Dover Publications.
National Fisherman
  1981 "Management Problems Hound Alaska's Troll Fishery," vol. 61, no. 12
        (April):8, 102.
Natural Resources Consultants
  1986 *Commercial Fishing and the State of Washington*. Seattle.
Netboy, Anthony
  1980 *The Columbia River Salmon and Steelhead Trout*. Seattle: University of
        Washington Press.
Northwest Indian Fisheries Commission
  1980 *Northwest Indian Fisheries Commission*. Olympia, Washington.
Northwest Indian News
  1979 "Full Text: Supreme Court on Boldt," vol. 10, no. 1 (August).
Norton, Helen Hyatt
  1985 "Women and Resources of the Northwest Coast: Documentation from the
        18th and Early 19th Centuries." Ph.D. dissertation, University of Wash-
        ington.

Orrick, Judge William H., Jr.
  1980   *United States* v. *Washington* (Phase II). U.S. District Court, Western District of Washington at Tacoma. Civil no. 9213. *Federal Reporter*, 506F. Supp. 187.
Ortner, Sherry B.
  1974   Is Female to Male as Nature is to Culture? In *Women, Culture and Society*, eds. Michelle Z. Rosaldo and Louise Lamphere, pp. 67–89. Stanford: Stanford University Press.
Plicher, William
  1972   *The Portland Longshoremen: A Dispersed Urban Community*. New York: Holt, Rinehart and Winston.
Porter, Russell
  1977   *Factors Affecting the Columbia River Chinook and Coho Salmon Resources in Fresh Water*. Reference document prepared for Comprehensive Salmon Management Plan of the Pacific Fishery Management Council. Portland, Oregon: Pacific Marine Fisheries Commission.
Rand McNally
  1984   *The Great Geographical Atlas*. Chicago: Rand McNally.
Rosaldo, Michelle Zimbalist
  1974   Women, Culture and Society: a Theoretical Overview. In *Women, Culture and Society*, eds. Michelle Z. Rosaldo and Louise Lamphere, pp. 17–42. Stanford: Stanford University Press.
Rudolph, Glenn
  1981   "Boldt's Decision Took Root at Medicine Creek in 1854." *National Fisherman*, vol. 62, no. 1 (May):38–40.
Seattle Times
  1982   "Offer Made to Buy Back Fishing Permits." January 3:C3.
Stansby, Maurice E.
  1963   *Industrial Fishery Technology*. New York: Reinhold Publishing.
Stevenson, Charles H.
  1899   "The Preservation of Fishery Products for Food." Extracted from *U.S. Fish Commission Bulletin for 1898*, pp. 335–563. Washington, D.C.: U.S. Government Printing Office.
Thompson, Paul
  1985   Women in the Fishing: The Roots of Power between the Sexes. *Comparative Studies in Society and History* 27:3–32.
Times Books
  1983   *The Times Atlas of the World, Comprehensive Edition*. New York: Times Books.
U.S. Bureau of Indian Affairs
  1977   *Background Information on Indian Fishing Rights in the Pacific Northwest*. Prepared by Bureau of Indian Affairs in cooperation with U.S. Fish and Wildlife Service and U.S. Department of the Interior, Portland, Oregon. Revised August 1979 by the Northwest Indian Fisheries Commission.
U.S. Department of Commerce
  1982   *Calendar Year 1981 Report on the Implementation of the Magnuson Fishery Conservation and Management Act of 1976*. National Oceanic and Atmo-

spheric Administration, National Marine Fisheries Service: Washington, D.C.

U.S. Department of the Navy
    1974 *Foreign Fishing, International Fishery Arrangements and Related Research Programs in the Eastern Pacific Ocean.* FISHACTS I, II, III. May 1971–April 1974.

U.S. Court of Appeals, Ninth Circuit
    1985 *United States v. Washington* (Phase II), Civil no. 81–3111. *Federal Reporter,* 759 F. 2d 135 (1985).

U.S. District Court, Western District of Washington at Tacoma.
    1974 *United States v. Washington* (Phase I). Civil no. 9213. *Federal Reporter,* 384 F. Supp. 312. (*See* Boldt, Judge George H.)

U.S. District Court, Western District of Washington at Tacoma.
    1980 *United States v. Washington* (Phase II). Civil no. 9213. *Federal Reporter,* 506 F. Supp. 187. (*See* Orrick, Judge William H., Jr.)

U.S. Supreme Court
    1979 Majority Decision. In "Full Text: Supreme Court on Boldt," *Northwest Indian News,* vol. 10, no. 1 (August):11–18. (*See Northwest Indian News.*)

Walsh, James P.
    1981 Developing and Implementing the Fishery Conservation and Management Act. In *Science, Politics and Fishing: A Series of Lectures,* ed. Naomi Krant, pp. 129–36. Corvallis: Sea Grant College Program, Oregon State University.

Wasserman, Jim.
    1980 "The Wheelhouse Revolution." *The Fishermen's News,* vol. 36, no. 20 (October):32.

Watson, Lawrence C., and Maria-Barbara Watson-Franke.
    1985 *Interpreting Life Histories: An Anthropological Inquiry.* New Brunswick: Rutgers University Press.

Williams, Woodbridge.
    1945. "King Crab—A New Fishery?" *Alaska Life* 8(2):22–29.

Index